EVOLUTION
OF STARS
AND GALAXIES

WALTER BAADE

EVOLUTION
OF STARS
AND GALAXIES

Edited by Cecilia Payne-Gaposchkin

The MIT Press
Cambridge, Massachusetts, and London, England

First MIT Press paperback edition, 1975
Original edition published by Harvard University Press.
©Copyright 1963 by the President and Fellows of Harvard College
All rights reserved
Printed in the United States of America
Library of Congress catalog card number 75-24679
ISBN 0 262 52033 8 (paperback)

EDITOR'S FOREWORD

W H E N Dr. Baade gave a series of lectures at Harvard Observatory in the fall of 1958, he hoped to embody them in a book. Because of his absorption in his work, and his well-known reluctance to write for publication, we arranged that the lectures should be recorded on tape. The recordings and the transcriptions were intended to serve him as a first draft. The transcriptions were a verbatim record of the lectures, including comments and questions by students and staff, which often led to spirited arguments in which no holds were barred. At Dr. Baade's request, copies of the transcriptions were given only to those who had been present at the lectures.

At the time of Dr. Baade's unexpected death, no progress had been made on the book, and there were requests from all parts of the world for copies of the lectures. As they stood, the transcriptions were unsuitable for publication. The colloquial expressions that give force to a speech have not the same effect on paper. But the pith of the lectures contains something permanent and irreplaceable — the final thoughts of a great observer on the subject to which he had devoted four decades of research.

In preparing a book on the basis of the lectures I have undertaken a great responsibility. Two qualifications I have: I attended all the lectures myself, and I have been concerned with the problems discussed in them almost as long as Dr. Baade was. In preparing the text I have tried to give precise expression to what I believe the lecturer intended to say, eliminating repetitions and correcting an occasional verbal slip. I could not hope to render the lecturer's inimitable and racy style, but those who knew him will catch an occasional glimpse of his explosive vehemence.

No attempt has been made to reproduce the discussion, but some comments by the lecturer that arose from it have been embodied in the text. The remarks that were made from the floor were "off the record": many were irrelevant, and the authors of others might not wish them perpetuated. The present book is intended to contain Dr. Baade's ideas, not those of his hearers.

Dr. Baade's friends know that he was dogmatic and emphatic, but they also know that he was not afraid to change his mind. What he said at Harvard represented his opinion in the fall of 1958; earlier he may have been of another opinion, and he would certainly have modified some of his ideas in view of what has been published in the interval. I have found it impossible to exclude his comments on unpublished results of which other astronomers had told him; if he misunderstood or misremembered these things, I hope that I shall not be blamed for setting down what he said, as I understood it. That I do not myself agree with all the opinions expressed in the book is true, but not important.

C. P.-G.

ACKNOWLEDGMENTS

FIGURE 1 is reproduced by permission of the Yale University Press. Figures 3, 4, 5, 6, and 18 are reproduced by permission of the University of Chicago Press, copyright © 1955, 1956, 1957, by the University of Chicago. Figures 7–17, 19, 22, and 25 are reproduced by permission from the *Handbuch der Physik*, vol. 51 (Springer-Verlag, Berlin, Göttingen, Heidelberg, 1958).

CONTENTS

TABLES

FIGURES

CHAPTER 1

HISTORICAL INTRODUCTION

The study of the evolution of stars and galaxies is in a stage of very rapid development, and we are still at the beginning. It is well to see the outlines of the field, and I shall begin with a very rapid survey. If I review some of the early results, we shall get the proper perspective, and can avoid dealing with them later.

The problem of the structure of the Galaxy is one of the most difficult that astronomers have encountered, and it is very important, because in our studies of other galaxies we must always fall back on our own for details. It goes back to the days of Sir William Herschel, who pictured a bilobed Galaxy simply by making counts of stars. We know today that the apparent bifurcation is caused by dark obscuring clouds that extend from Cygnus to southern latitudes.

The nineteenth century was the time of the collection of data, especially on the motions of the stars. Only in the

1840's, when Bessel measured the first trigonometric parallax, did information become available on the distances of stars. But by the end of the nineteenth century so many proper motions had been measured that the luminosities of groups of stars could be derived from parallactic motions. At the same time radial velocities became available.

By the beginning of the present century it had become obvious that the stars have a large range of luminosity. The number of stars of a given luminosity in a given volume of space (the so-called luminosity function) was derived by Kapteyn and other astronomers, and was a very ingenious piece of work. A new phase of the problem was the realization that the luminosity function was connected with the number of stars seen along a given radius (the so-called density function) by an integral equation. The problem of determining the structure of the Galaxy now involved the study of the distribution of stars in all directions around the sun.

The new approach appealed to the mathematicians, and some of them, like Karl Schwarzschild, gave elegant solutions for solving the problem numerically. But astronomically it was a nightmare; one could predict that generations of astronomers would go to their graves before seeing any results; and the results, when they came, would at best be hazy. Kapteyn himself constructed a first approximation to the picture: the famous "Kapteyn Universe," with the sun in the center, a flattened system with axial ratio about 1:4, concentrated toward the plane of the Galaxy and toward the center. The smaller axis was about 1500 light years, the larger, about 6000.

Today nobody would use the luminosity function to probe the structure of the Galaxy. Currently the luminosity function interests us for a very different reason: it gives valuable information about the history of star formation in the Galaxy, as we shall see later.

By 1900 there were three new developments that led directly into the present stream of our thinking. The first was the determination of parallaxes with large photographic refractors, which was developed by Schlesinger. This work is usually taken for granted, but it is fundamental to the whole of modern astronomy. We now have excellent trigonometric parallaxes for stars within 10 pc, and these stars furnish very important criteria of the correctness of our conclusions.

The second development, which was to play just as big a role, was the survey of stellar spectra. By 1900 two of the Harvard catalogues had been completed — Miss Cannon's for the southern sky, and for the northern stars Miss Maury's famous catalogue, which Hertzsprung immediately put to a most ingenious use, as I shall mention later.

The third fundamental advance was the old Boss catalogue of proper motions of high precision; it was just a fantastic achievement, which has never been surpassed. Its importance was recognized immediately, and all the great astronomers of the time — Eddington, Schwarzschild, Hertzsprung — obtained valuable results from it.

The next great step, the Hertzsprung-Russell diagram, grew out of the new material on parallaxes, proper motions, and spectra. Without these it would have been impossible. Around 1903 Hertzsprung began to publish his studies of selected groups of stars in a series of papers that are still relevant today. For the first time it was possible to pick out groups of similar stars, and here Hertzsprung was unsurpassed. One of the first groups that he studied consisted of what we now call the high-luminosity O and B stars. The Boss catalogue gave the precision proper motions, and the Yerkes Observatory provided radial velocities for most of them. Hertzsprung noticed especially the stars with exceedingly sharp lines that Miss Maury had called c-stars in her catalogue, and he showed that they were stars of very high luminosity, -5^M in the mean. He tried to determine the dispersion, and showed that some of these stars could be as bright as -7^M or -8^M. As he continued the investigation of c-stars of other spectral types, he found that they all had exceedingly high luminosities; today we call them supergiants. All this was already done in 1903. The papers that Hertzsprung published between 1903 and 1910 are fantastically modern; he showed for the first time that there is a group of stars of very high luminosity.

For the stars of lower luminosity it was easy to get information from the growing list of trigonometric parallaxes. Here for the first time Hertzsprung noticed a systematic division of stars of later type, beginning at G, into groups that were to be called giants and dwarfs.

These conclusions attained general recognition when Russell published them in 1914 in the form of a diagram. He plotted spectral type against luminosity, and showed the existence of a main sequence and a giant sequence; the number of known giants was very small. The work grew out of his interest in eclipsing systems, and the bright components of a number of these fell on the giant branch, and among the very rare supergiants.

I was a young student at the time, and the discovery of the Hertzsprung-Russell diagram (the H-R diagram) made a tremendous impression on me, because obviously it opened up an entirely new way of studying galactic structure. We could pick out well-defined types of stars with relatively little dispersion in absolute luminosity, and use them to analyze the structure of the Galaxy.

The H-R diagram was at once recognized as an exceedingly useful tool for determining the distances of galactic clusters. The outstanding work at that time was Hertzsprung's great study of the Pleiades; he plotted his own magnitudes against the spectra down to the tenth magnitude. He also determined cluster membership very carefully by means of proper motions; the Pleiades is one of the few clusters for which this can be done. He went further: because of the difficulty of obtaining spectra of faint stars, he replaced spectral type by color index, a property that had already been suggested by Schwarzschild. Besides plotting luminosity against spectrum, Hertzsprung plotted luminosity against a measure of color index that he called "effective wavelength," and thus determined the color-magnitude diagram of the Pleiades from the second to about the fifteenth magnitude. In cooperation with Seares he investigated other clusters, including NGC 1647.

But this development soon came to an end. All the photometry had to be done photographically, either from a comparison of photographic and photovisual magnitudes, or from the determination of effective wavelengths, which were of rather low precision. Standard magnitudes were available in the North Polar Sequence, but the magnitude scales were very difficult to transfer, and scale errors could easily affect the color indices. There were soon dozens of papers on the color-magnitude diagrams of open clusters, but the discrepancies discouraged most of the workers, and the studies were not continued.

There was one exception. At this time Trumpler began his grand attempt to study the open clusters of the Galaxy. He determined the spectra of the brighter stars, and approached the color indices of the fainter ones with infinite pains. He did the best that could have been done before the era of the photocell. He determined not only spectra and color indices, but also radial velocities. He died before the work was finished; when his last papers are published they will furnish important data on the kinematics of clusters.

One of Trumpler's big discoveries was the existence of interstellar reddening and absorption. The problem of absorption was an old one; astronomers of the preceding era, Kapteyn and Seeliger, had worked out the formulas, but no observational evidence had been produced. Trumpler found it by the remarkable and simple procedure of grouping together clusters of the same stellar content, and seeing whether their numbers increased with distance in the expected way. He found that they did not, that absorption was present, and that this absorption was accompanied by reddening. When the photocell came into operation in the 1930's this lead was immediately followed up.

On account of the difficulty of transferring magnitude scales, the study of the H-R diagram ran into trouble. Another method, which was initiated 25 years ago by Lindblad, will certainly become important with the aid of the photocell; it is ideal for the purpose. Lindblad used objective-prism spectra to determine luminosity criteria with the aid of selected stars of known luminosity. In his very first papers he showed that it was possible to determine the luminosities of A and B stars from the strength of the hydrogen lines. For stars of later type he used other criteria, such as the strength of the G band, and he was the first to point out that the strength of the cyanogen bands varies with luminosity. He measured the continuum of A and B stars, and by comparing with the photographic magnitudes he determined color indices. He even tried to measure the reddening; by comparing color with spectral type he found the color excess. So he had all the data necessary for a statistical investigation of luminosities. He and his collaborators worked with small instruments, but I am very certain that his method will come into its own now that more powerful instruments are available, especially with the new technique of photoelectric scanning.

The H-R diagram had opened up an entirely new way of approaching the problem of galactic structure. At the same time, early in the century, another very interesting series of developments provided us with a powerful tool. At the Harvard Observatory Bailey began his study of variable stars in globular clusters, and at the same time Miss Leavitt began the investigation of the Magellanic Clouds. Bailey found that the globular clusters are full of RR Lyrae stars. At that time only a few RR Lyrae stars were known, scattered over the whole sky, but when Bailey had finished his work he had found more in globular clusters than were known everywhere else.

The vast majority of the variable stars found by Miss Leavitt in the Magellanic Clouds turned out to be Cepheids, and she found when she plotted apparent magnitude against period that the shorter the period, the fainter the Cepheid. Since it was possible to suppose that all these stars were at the same distance, it was reasonable to conclude that period is actually related to luminosity. The work of Bailey and Miss Leavitt had profound influence on what was to follow.

Hertzsprung had already studied the Cepheids of the Galaxy, and he immediately used Miss Leavitt's results for his famous paper on the period-luminosity relation, which appeared in 1913. The way was opened up for the determination of distances by means of Cepheid variables.

The first extensive use of variable stars in the determination of distances, however, rested not on the Cepheids but on the cluster-type variables. This was the very remarkable work of Shapley, based on Bailey's study of these variables. At the time these stars were simply thought to be Cepheids with periods less than a day. The idea seemed plausible, and it was justified by the fact that the light curves and radial velocities of the two brightest cluster-type stars were related in the same way as they are for the Cepheids. So there was a chance to use these stars to determine the distances of globular clusters.

Hertzsprung had determined a period-luminosity curve for the Magellanic Clouds, where either there were no cluster-type variables or the cluster-type variables could not be reached. How, then, could the luminosity of these stars be determined? Shapley's very ingenious idea was to use the few Cepheids that Bailey had found in globular clusters. In ω Centauri Bailey had found five or six such stars in addition to the numerous cluster-type stars; the latter had all the same luminosity, independent of period. When Shapley made a period-luminosity relation for the Cepheids in ω Centauri he found that its slope closely matched that for the Cepheids in the Magellanic Clouds. So, with the same absolute magnitudes, he derived an absolute magnitude of $-0^M.23$ for the cluster-type variables. As we shall see later, it was not correct to use a unified period-luminosity relation, but curiously enough the deduced absolute magnitude for the cluster-type variables was very nearly correct, and Shapley's results for the system of globular clusters have been unaffected by the later changes in the period-luminosity curve for the Cepheids.

The globular clusters presented an intriguing problem. In 1913 Hinks had pointed out the very remarkable distribution of the globular clusters in the sky, as others, indeed, had done 40 years earlier: almost all of them are in one hemisphere. What Hinks had suspected, that the center of the Galaxy lies in the direction of Sagittarius, was shown by Shapley's most remarkable work at Mount Wilson. I have always admired the way in which Shapley finished this whole problem in a very short time, ending up with a picture of the Galaxy that just about smashed up all the old school's ideas about galactic dimensions. He showed that the center of the Galaxy, at the center of the system of globular clusters, is at a distance of 10 kpc. This was before Trumpler's work, and took no account of absorption, but it is remarkable that the distance was of the right order.

It was a very exciting time, for these distances seemed to be fantastically large, and the "old boys" did not take them sitting down. But Shapley's determination of the distances of the globular clusters simply demanded these larger dimensions. The Galaxy now appeared as a very extended lens, and we had every reason to believe that our sun was some 10 kpc from the center. It was surrounded by a spherical system of globular clusters. The picture was still vague, but for the first time it had some outlines. And in some respects the outline was filled in very rapidly by the discovery of many isolated cluster-type variables in the spherical system: besides the globular clusters we had a galactic halo. The number of known cluster-type variables was soon so large that the idea of their being escapees from globular clusters became untenable; it would have required eight or ten times as many globular clusters as were known at the time.

In connection with his study of globular clusters, Shapley had for the first time determined color-magnitude diagrams for three or four of them. He clearly recognized that these diagrams were entirely different from the standard H-R diagram, even though he was limited to about the seventeenth magnitude by the lack of fainter standards. But at the time there was no sound basis for understanding the difference.

In succession to this work of Shapley's, the problem of the extragalactic nebulae was tackled a few years later by Hubble, who showed conclusively that the Andromeda Nebula and others are galaxies like our own. From the beginning, Hubble had to fight a tremendous handicap. Shapley, in his work on globular clusters, was dealing with objects so bright that they could be handled with the available photographic standards, which were reliable down to the seventeenth magnitude. But in his work on extragalactic nebulae, Hubble needed magnitudes around 18, 19, and 20, and these magnitudes had to be extrapolated. So one of the problems of the time was the establishment of proper magnitude scales, a problem that we are only now overcoming, with the aid of the photocell. And it is a very difficult job, even for the photocell. I should like to stress the very fundamental importance of the problem of magnitude scales.

CHAPTER 2

CLASSIFICATION OF GALAXIES

T H E descriptions of galaxies in the NGC, which go back to the Herschels, refer essentially to the sizes and give rough indications of brightness. More could hardly be expected, because these early observations were all visual, and everybody who has looked at galaxies like the Andromeda spiral knows that the visual pictures are lacking in detail. Actually it is only a hundred years since Lord Rosse, with his big reflector, found spiral structure in a number of galaxies.

The situation changed when visual observations were replaced by photography at the turn of the century, particularly when the large reflectors came into operation — first the Crossley reflector at Lick Observatory, later on the Mount Wilson reflectors. Pictures of the nearer galaxies showed such a wealth of detail that it was necessary to bring some order into the observed forms and structures.

Some early classification schemes, such as that of Max

Wolf, were based on photographs made with astrographs of intermediate size, but they did not find much acceptance. In the early 1920's, two classification schemes were proposed almost simultaneously, one by Lundmark, the other by Hubble. Both were based on the large collections of plates that had been made at the Lick and Mount Wilson Observatories. There was not much difference between the two systems. Lundmark tried to classify many details in addition to the main features. It so happened that in the long run the simpler classification of Hubble won out, and it has been in general use since that time.

Let me remind you of the essential features of Hubble's system. He started with a class that he called the E nebulae — the spheroidal galaxies. They·run from round forms ($E0$) to elliptical forms ($E5$). We can leave out $E7$, which Hubble classified in an entirely different way, and $S0$, about which I shall speak later. The letter E stands for elliptical, and the numeral for the ellipticity, simply defined by the ratio $(a - b)/a$, where a is the major axis and b the minor axis in the isophotes. I talked with Hubble about the whole thing a few years ago, just before he died, and it seemed that he was convinced that the limiting case was about $E6$, with the smaller axis about 40 percent of the larger axis.

The E galaxies are distinguished by a remarkably smooth intensity distribution, as we know from the early work of Hubble and the later work of de Vaucouleurs and others. Many of these galaxies have a very remarkable feature — a very bright semistellar nucleus at the center. The companion of the Andromeda spiral, M 32, furnishes one of the brightest cases in which this nucleus can easily be seen.

We shall see later that a central nucleus is also a feature of the spirals; though sometimes exceedingly weak or even absent, it is present in many cases. If it is present in spirals, which are flattened systems, it shows flattening. For instance, the central nucleus of the Andromeda spiral has dimensions about 2″.5 × 1″.5, and its axis is directed beautifully in the direction of the axis of the Andromeda spiral itself. So this nucleus, this high condensation at the center, is a feature of all these galaxies.

Here we have a very remarkable thing. We are dealing essentially with two systems — a central spherical system, surrounded by a disk that exhibits spiral structure. In a spiral galaxy, the central system (especially evident in NGC 5494, for example) can shrink until it is finally reduced to a semistellar point. A very short exposure would be needed to show the semistellar point in M 101; what is visible in the photograph is a much larger bright area.

In classifying the spirals, Hubble distinguished the groups *Sa, Sb* and *Sc,* the distinguishing criterion being essentially the behavior of the spiral arms. For instance, in his description of *Sa* the spiral arms emerge at the edge of the central system; in the early spirals both they and the central lens are still unresolved, and the arms are densely coiled. As we proceed along the series, the central nuclear area shrinks at the expense of the growth of the arms, which by and by uncoil, until finally the central area has shrunk to a semistellar point, and all the mass seems to be in the spiral arms. This is Hubble's original description, but he agreed completely that it would be simpler today to classify the spirals simply by the size of the central lens. Those with very large central lenses can be called *Sa;* those with intermediate central lenses, *Sb;* and those where the lens has shrunk to a semistellar point (actually a huge cluster of stars), *Sc.* In what follows I shall adopt this very simplified system, based on the size of the central lens.

Parallel to this series of ordinary spirals is the series of so-called barred spirals, which typically have a central system — a bar within a ring area — and spiral arms springing from this ring. In Hubble's classification, *SBa* is a type that is crossed by a bar, and the rest of the material is in the form of a ring around it; this ring breaks up at two points, and begins to show spiral structure. In the final stage the bar shrinks at the expense of the finer spiral arms, and we then find the latest type of barred spiral, *SBc*.

Whereas all the systems discussed so far show obvious rotational symmetry, there were left about 2 to 3 percent of the galaxies in which rotational symmetry was not obvious. They were irregular in shape, and Hubble classified them as irregular galaxies. Today we know that they are exclusively of the Magellanic Cloud type.

The final form in which Hubble presented his sequence is shown in Fig. 1. He started with *E*o ; *E*6 is a limiting case

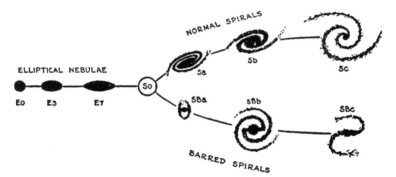

FIG. 1. The sequence of nebular types.

for elliptical galaxies, and beyond that there was a bifurcation into two parallel series, the normal spirals and the barred spirals, with some intermediate cases in between. At the junction he postulated a hypothetical class *So*, which marks the transition from the *E* galaxies to the spirals.

In his first paper Hubble admitted that, although he could perhaps find transitional cases between the *E* and *SB* series, he had not yet found a single case of spiral systems without spiral arms. Later on, such systems were actually discovered — systems where spiral arms would be expected but were not found; Hubble distinguished such systems as *So*. Such galaxies have a central lens — a disk — but no spiral structure. We know today that *So* systems are the prevalent type in dense clusters of galaxies, and, besides showing no spiral structure, they show no trace of gas or anything of the sort. The explanation of these systems is entirely different from what Hubble thought it was. We know today that the reason is that the dust has been wiped out by collisions between the cluster galaxies. We shall see later on that this is not true for the field galaxies of type *So*. But it has nothing to do with evolution.

In his latest classification (which is given in the *Hubble Atlas*, edited by Sandage) Hubble changed the position of *So* and, as we shall see later, put in its place a new series of stripped spirals, the *So*'s: *Soa, Sob, Soc*, parallel to *Sa, Sb, Sc*. This *So* series has been stripped of gas, or (as in some cases that we shall see later) the gas has been exhausted. This series is devoid of dust and gas, and shows no spiral structure. You would get exactly such systems if you were to strip the spiral structure from any of the great spirals.

The irregular systems show at first sight very little indication of rotational symmetry. But for a number of reasons the appearance is deceptive. There are clear indications that these systems are very flat and in a state of rotation. When we come to IC 1613 we shall see that it is a very irregular galaxy with a most ragged outline, but still, when the Population II is traced out with the radio telescope, it is found to have a perfectly beautiful elliptical outline.

The ragged outline shown by the photographs simply indicates the area in which I think star formation is at present going on; the mass distribution is absolutely regular.

Another argument relates to the Magellanic Cloud systems, which we see in all orientations in the sky. When we see them sideways they are always highly flattened. But when we see them face on, or very nearly face on, like IC 1613, we observe no concentration toward the center, or only a very slight one; if they had any other form we should observe it. Therefore they are highly flattened systems. Finally I may add that a regular rotation of the Magellanic Clouds has been found from both optical and radio observations.

Magellanic systems very often show a central bar. These systems may be closely related to the barred spirals. We practically always find one bar, and we must be cautious in calling these systems irregular. The outline looks irregular, but the chaotic outline is that of the present area of star formation. I should like to set the matter of the irregular galaxies right at this point: we have every reason to believe that they are highly flattened systems, though not as flattened as *Sb* or *Sc* spirals, and in consequence are probably in a state of rotation.

The irregular systems cannot be very young, since they contain Population II stars and cluster-type variables. The irregular features consist of young stars, supergiants and so forth, which make a terrific splash and contribute lots of light. But their total mass can be very small compared to the total mass of the system.

Now it is obvious from the scheme as Hubble described it that he had an impression or a belief, although he never quite admitted it, that it represented a continuous sequence. But I believe, on the contrary, that Lindblad put his finger on the essence of Hubble's classification when he suggested that it is a series of increasing flattening, of increasing angular momentum.

It is a very remarkable thing that in the E series we have very simple systems from $E0$ to $E6$, and among the spirals we have in general two systems, one of high angular momentum (the disk), and one of low angular momentum (the lens), for a system of the same mass.

Even if we leave out considerations of angular momentum, we have a series of increasing flattening. This certainly presents a problem, a very important problem. Aside from the E galaxies, why do the spirals always show the combination of a disk and a central spheroidal system? It must reflect the original density distribution of the gas. It is hardly possible to think anything else, because otherwise we simply could not understand the great irregularities of the density distribution of subsystems, which are well known; they could not even have established themselves.

This is really a very serious problem for the theoreticians to look into. It almost seems as if, in the spirals, most of the original angular momentum has gone into the disk. Can we imagine that, at some era in the past, the central spheroidal system of low rotation and the disk with very fast rotation actually resembled the equilibrium figure of the gas? One should really look into these things.

To return to the Hubble classification: of course his catalogue referred to the brightest galaxies; there was a selection according to luminosity. His classification of the E's has to be extended because of the discovery in recent years of E galaxies of very small size and low luminosity, of the type of the Sculptor and Fornax systems. When they were discovered at Harvard, Shapley regarded them as a new type of stellar system. We know today that they are simply the continuation of the ordinary series of *E* galaxies, which starts from giants like M 87 in the Virgo cluster or M 32; they represent the smallest *E* galaxies that we know today. Examples are the Leo II system, a spherical galaxy about the same size as the Sculptor system, containing over 300 cluster type variables; the Leo I system, an elliptical galaxy near Regulus; and the elliptical Draco system. Such galaxies must in future be covered by the classification system. There is a whole series of luminosities, and as the luminosity decreases, the pictures look very different from the beautiful standard *E* galaxies in Hubble's original paper.

Let me add a few words about the merits of Hubble's system. It is a very simple one, but we shall see later, when we try to understand the make-up of galaxies, that there is really not much sense in making a system that covers all the little details of spiral structure. About the merits of Hubble's system I can speak from experience; I have used it for 30 years, and, although I have searched obstinately for systems that do not fit it, the number of such systems that I finally found — systems that really present difficulties to Hubble's classification — is so small that I can count it on the fingers of my hand.

Actually, Hubble's system is even better than one would expect from his own classifications. In his work on the Shapley-Ames catalogue he often made the remark "peculiar," as if his system was not a good one. The reason for this "peculiar" was that Hubble had a blind spot where two galaxies were involved instead of one, a case that happens quite frequently. I remember the difficulty that I had in convincing him that a certain galaxy is double. When we finally showed that the two systems had different radial velocities, he still called it a hypothesis. If you eliminate the double systems, I am sure that the number of exceptions is unbelievably small, so efficient is the system.

A good system should be applicable over the whole observed range. Every time I took long exposures with the 100-inch or 200-inch (90 minutes or so in the blue, about 4 hours in the red) under excellent conditions, I always examined the numerous small and distant galaxies distributed over the field to see how they behaved and whether they followed the classification scheme, even when the plates were taken for another purpose. Even systems with diameters of 5″ could easily be classified, especially with the red and blue plates put together in the blink microscope. The *Sa, Sb,* and *E* galaxies were easy. The more difficult cases were the *Sc*'s, where the central nucleus has shrunk to a semistellar point; on the blue plates one very often saw only a little elliptical patch. But when one switched to the red image, the central nucleus was visible, and one was absolutely certain that one was dealing with an *Sc.*

I think that the Hubble classification, which just deals with the basic features, conveys all that we want to know. It will become clear later that all the details of the spiral structure are what might be called accidental — variations on the same theme. I believe that the more recent systems, like those of de Vaucouleurs, are a retrogression to the time of Lundmark, and even beyond. I can see no sense in covering these details; even if you put them all in you still do not get the full picture. If you want to study the variations on the theme *Sc,* you simply have to take the plates and examine them — only then do you get the full story. No code system can replace this. The code system finally becomes so complicated that only direct inspection of the plates helps.

I think that the search for exceptions to Hubble's classification is very significant. One of the examples that I said I could count on the fingers of one hand is IC 51. I found it on Schmidt plates and could not make head or tail of it. Then I took a photograph with the 200-inch. The system has a very regular disk, but there is a remarkable thing: smoke rings come out of the nuclear region. If it were not for this feature, it would be an *Sb-Sc*. It is not a serious exception. In another case, although the system shows central symmetry, the dust pattern violates the symmetry completely, shooting out to large distances, although there are no external systems in the neighborhood. If we were dealing with two systems that had totally collided, on a simple expansion picture with the velocity of separation we should find the culprit within a few degrees, even for the brightest galaxies. In this case nothing was to be found; it is a real exception.

The irregulars were the wastebasket in Hubble's system, because he dumped his doubles into this class. But he still stated that the greater part of the irregulars were of the Magellanic Cloud type. We know now that they are exclusively of the Magellanic Cloud type. We shall see later that there is a group among the irregulars that do not show any Population I stars (giants and so forth); M 82 is one of these. It is remarkable, with color index about +0.9 and spectral type A5. Holmberg has distinguished Irregular I (where Population I dominates) and Irregular II (where Population II dominates) on the basis of color index. Hubble's classification of irregulars by forms is obviously a good thing; and with the forms alone we get a very close correlation with content.

The distribution of galaxies in space is very chaotic, and far from uniform; single systems are rather rare. There is a story here: Hubble and I had a long-standing bet of $20 for the one who could first convince the other that a system which he had found was single. We never could decide the bet; neither of us could pull out some distant fellow — in some cases there really was a companion and in other cases there could be. So single galaxies may be rare. Doubles and multiples are frequent; from these we go to groups, from a dozen to a few hundred; from there to clusters. In the Coma cluster and others we even see a strong concentration toward the center. Finally we come to clouds of galaxies, which are usually not mentioned. At the present time we can draw no sharp distinction between these different groupings.

If you stagger all these clouds through space, only a few of the nearer ones will stick out. Earlier, Shapley at Harvard found some. One of the most amazing of these clouds is the one found by Shane in his Lick Survey at about 15^h 20^m, $+5°$; the globular cluster M 5 lies at the edge of this cloud. The Lick Survey gives some 200 galaxies per square degree. At Shane's suggestion I had plates taken a number of years ago in the blue and the red with the 48-inch Schmidt, which is much more powerful than the 20-inch. It is really most fantastic; the area is literally covered with galaxies. And the interesting thing is that on this plate you find clusters of galaxies dispersed everywhere in this large cloud, which means that a large cloud contains a large number of clusters of galaxies. Another interesting point is that the galaxies of the cloud are preferentially *Sb* and *Sc*, but in the clusters themselves we find *So*'s, although all of them are obviously connected with the cloud.

We do not know the borderlines of the subdivisions. Holmberg's beautiful paper in the *Lund Publications* refers only to the doubles among the brighter galaxies of the NGC; it is a model of reasoning. Among the groups we have Stefan's Quintet, and the one that Seyfert found a few years ago; you find these groups everywhere. If you look at the Palomar Sky Survey plates they are unavoidable. You see them together in relation to their types; you know it is a unit; you can check on the radial velocities. But you can only divide according to numbers after they are better understood.

If we restrict the term "clusters" to the very rich associations of galaxies which at the same time show a strong concentration toward the center, like the Coma and Corona Borealis clusters, then we should have to include the Virgo "cluster" among the "groups." So the "groups" at the present time are simply a catchall for intermediate concentrations, from multiples on; they would contain anything from half a dozen, eight, ten galaxies to several hundred, as in the Virgo cluster. It remains to be investigated whether the larger groups, such as that in Virgo, should be regarded as clusters.

It is very simple to decide whether a star is a member of a star cluster; because the total mass is very small, one can always fall back on proper motion and radial velocity and can always decide about membership. In the case of clusters and large groups of galaxies, the dispersion in velocity is very large, so the red shift does not mean anything. It is very hard to assign individual members even to a cluster of galaxies, and in the case of the Virgo group the problem is truly difficult. It is quite easy to derive the brighter part of the luminosity function for clusters like the Virgo and Coma clusters. But it is practically impossible to get the fainter end, because we are counting against a rising background; as we come to the faint end we are simply lost statistically, and have to use physical arguments. We shall find that such physical arguments can eventually enable us to disentangle these problems, but it has not yet been done.

The problem of the stability of the groups is analogous to that for associations of stars in our Galaxy. The groups with positive energy will long ago have dispersed, and we may call semistable the groups that are left now. Occasionally, owing to a close encounter, a member gains sufficient energy to be ejected, but we are certain that the chance for a capture is negligible in the present state of the expansion of the universe, and we can forget about it. And it is an important corollary that the members of the groups that we observe today are of common origin. This will be especially important later on in connection with some conclusions that we shall draw regarding our own group of galaxies: they are all of common origin, and therefore we have good reason to believe that as galaxies they are of the same age.

Our own Local Group may consist of 17 or 18 members. Among the next brightest galaxies, NGC 2403 is the main member of a group; M 81 is the main member of a brighter group, probably just as extensive as our own, because we know we have not found its faintest members like our Sculptor system. Third comes M 101, where again we are in an extensive group, all belonging together. And there are dwarf systems marching in between, such as the Sextans and Wolf-Lundmark system. So in our immediate neighborhood we have the typical arrangement — all the systems in dense groups, and these groups part of a larger one.

We are fortunate that our Galaxy is a member of one of these groups, the so-called Local Group. In Table 1 I have arranged them according to luminosity. The system of highest luminosity (about −20 photographic, actually

Table 1.   Local group.

Galaxy	Type	Absolute magnitude
M 31	Sb	−20
Galaxy	Sb	?
M 33	Sc	?
LMC	Irr	?
SMC	Irr	?
NGC 6822	Irr	?
IC 1613	Irr	−14
IC 5152	Irr	
M 32	E	−15.4
NGC 205	E	−15.1
NGC 185	E	−13.6
NGC 147	E	−13.3
Fornax	E	?
Leo I	E	−12.0
Sculptor	E	−9.5 to −11
Draco	E	−9.5 to −11
Leo II	E	−9.5 to −11
UMi	E	−9.5 to −11

a little brighter) is the Andromeda Nebula, M 31. Next comes our own galaxy, though of course we do not know the luminosity. The Large Magellanic Cloud is probably more luminous than M 33; NGC 6822 and IC 1613 are dwarf galaxies of the Magellanic Cloud type. A probable new member is IC 5152, a typical dwarf galaxy in the southern hemisphere recently photographed by Evans; from the reports and the picture I should guess that it is an actual member of our Local Group. Then come all the E galaxies. The series of spirals and irregulars seems to stop around −13^M. The ellipticals include the near companions of M 31, then NGC 185 and NGC 147, the more distant companions of M 31, then Fornax, absolute magnitude unknown; the last four in the table have about the same luminosity. The figures given are all on the correct new photometric scale, and will not be changed any more.

Leaving out the new member, our Local Group consists of 17 systems, of which 10 (59 percent) are E galaxies, of intermediate to faint luminosity. This picture is very different from what we get over the sky at large, where we give preference by selection to the brighter systems. The luminosity range of galaxies is from about -20 to about -9.5 or -10; this is a range of 10 magnitudes, a ratio of $10^4:1$.

The largest system in the Local Group, and one of the largest in the Universe, is the Andromeda spiral. If we transfer it to the Virgo cluster, whose distance is known from the red-shift –velocity relation, we see that there may still be a handful of systems in the Virgo cluster that are still brighter, but it is really near the top. Its absolute magnitude is -20, and — this is conservative — it has a linear diameter of about 50 kpc out to the distance of the large H II regions on either side. As an example of the smallest diameter (all these small ones are about the same) I give that of the Draco system, 1.6 kpc, determined from the most outlying cluster-type variables. So the ratio in diameters is 1:30. Thus there is a huge variety among galaxies in luminosities and in diameters. I need hardly say that the main mass of the Local Group is contained essentially in M 31 and our own Galaxy — or if you wish you can include the systems up to the Small Magellanic Cloud. If you add up the masses you come to something very close to 5×10^{11} solar masses.

A similar variety in size and luminosity is shown by M 31, M 32, IC 1613, NGC 185, and NGC 147, which are all practically at the same distance, so a comparison of their apparent dimensions gives a true picture.

It has been suggested that NGC 6946 and IC 10 may be members of the Local Group, but NGC 6946 is definitely not a member; the absorption is well determined and the red shift is too high. Its distance modulus on the new scale is 26.8 or 26.9. But IC 10 could be a member; it is one of the incredible cases that one would not believe possible. A photograph shows a beautiful piece of a spiral arm, with three H II regions of which we have taken the spectra. And that is the only thing you can see; you look everywhere around, and no extragalactic nebulae come through — there must be just a hole through which we see this piece of spiral arm. The radial velocity is not yet decisive, because the red shift depends very much on the assumptions one makes about the component of galactic rotation in that direction. To get the absorption an enthusiast, an optimist, would have to observe some of the blue stars in this system; but it could be done.

CHAPTER 3

THE OBSERVATION OF GALAXIES

T H E Local Group is our main field of exploration if we want to do detailed work on galaxies. In fact, if our own Galaxy were not a member of such a group, we should be at a very great disadvantage. The first investigations of the Local Group were made by Hubble and Shapley. After Shapley had made his investigations of globular clusters, he had really made a breakthrough as far as exploration of the galaxies was concerned. The period-luminosity relation for Cepheids had been available since 1913 in the form of absolute magnitudes. Shapley had to make a little detour to make it applicable to globular clusters, because globular clusters contain primarily cluster-type variables, which he had to tie up with Cepheids.

However, Shapley did not continue this investigation. When I met him in Hamburg in 1920, just after his work on globular clusters was finished, I asked him why he had not continued right away with the galaxies. Ritchey's picture of M 33, taken with long exposures at the 60-inch, had just been published, and it showed that the system was simply covered with stars, especially along the spiral arms. When I asked Shapley why he did not study the spiral form of M 33 he told me that the many images were not really stellar, that compared to stars they were "soft," and quoted Ritchey as authority. Shapley was so much impressed by this that, although I urged him to try it, he was not convinced and made no attempt to study M 33. Actually he must have been convinced at that time that the spirals were not extra-galactic, but belonged to our Galaxy; at about the same time, in the famous debate with Curtis before the National Academy of Sciences, Shapley defended this view. From the debate it is clear what his main arguments were.

The first argument rested on van Maanen's measures of the rotation of galaxies, but these had been done in the old-fashioned way with a micrometer, comparing old plates taken by Ritchey with plates that van Maanen had taken himself. Ritchey wanted to take very beautiful pictures of the sky, and for the first two or three years all the time of the 60-inch was reserved for him to take these plates. In order to get the most beautiful pictures possible, he used exceedingly fine-grained plates, which was excellent, but which necessitated exposure times of 6, 8, even 10 hours. Of course he occasionally ran into poor seeing, and, in order to avoid its effects, he drew the slide every time the seeing was bad and sat at the telescope until it got fine again. Now if you look at Ritchey's pictures of such a field with a low-power magnifier, the images indeed look magnificent, and beautifully round. But I once measured these plates in a test of van Maanen's results, and I found a very disturbing thing — the outer image is perfectly round, but the distribution inside the image is asymmetric. The center of density does not coincide with the center of the image; it simply refers to the part of the exposure time when the accumulation was highest. If you measure, as van Maanen measured, with a magnification up to 5, for the brighter stars you will simply set on the geometric center; the moment you come to fainter stars you will finally set on the center of density; that is one very disturbing effect. Another effect is produced by not very accurate centering in the field; over the field you always have residual coma, which leads to a slight effect that looks like rotation. So these rotations came partly from seeing errors, partly from centering errors, plus Ritchey's peculiar procedure.

Therefore when another man, like Lundmark, measured the same plates he simply got essentially the same results. In the middle 1930's Hubble had measured the plates and did not agree with van Maanen's very fine rotations. Nicholson and I were asked to remeasure these plates; I measured with high magnification, Nicholson with low. Nicholson again found something like rotation; I, like Hubble (who used large magnifications, so that he could always distinguish between the geometric center of the large image and the real center of density) got no rotations whatsoever. The problem is now completely cleared up. But the van Maanen rotations played a big role in the 1920's; Shapley was apparently thoroughly convinced by them.

The arguments given by Curtis for the extragalactic nature of the spiral nebulae were not too strong, and details were left out, but he pointed out that novae appeared in quite appreciable numbers in M 31, although not much was known about their luminosities. But Shapley attacked him with the argument that the scatter in the luminosities of these novae must be very high. The novae that had been found around 1920 were what we call today common novae, and in the Andromeda nebula they reached maxima of 15^m down to 18^m. But the supernova S Andromedae had also been found near the nucleus, of magnitude $7^m.5$. It was a confused situation; in order to show that it was impossible for the system to be outside the Galaxy, Shapley used the argument that S Andromedae would have the unbelievable absolute magnitude of -12 or -13 if one assumed the fainter novae there to be similar to our ordinary novae. He thought that no star could reach such a luminosity, but he did not publish a proof.

Another remarkable thing came up at about the same time. Why did nobody recognize that the Magellanic Clouds are the nearest extragalactic systems? In the 1870's, Cleveland Abbe (later Director of the U. S. Meteorological Survey, who earlier spent several years at Pulkovo Observatory) wrote a paper to whose conclusions we can subscribe today without any reservation. He showed that all the evidence (essentially only visual) available at that time indicated that the Magellanic Clouds are undoubtedly the nearest extragalactic systems. He took his evidence from the detailed descriptions by Sir John Herschel — stars, star clusters, emission nebulosities, and so on. Since the time of the Herschels the general idea had been that the Magellanic Clouds, because of their similar looks, were detached clouds of our own Galaxy.

In the heat of the discussion about the rotation of the galaxies, Abbe's work was forgotten. The next reference that I can find to the Magellanic Clouds is in the paper by Ralph Wilson in volume 13 of the *Lick Observatory Publications*. Wilson had observed all the available emission nebulae in the Magellanic Clouds, and pointed out that from the pictures and the other evidence they are most likely the nearest galaxies.

In about 1920 Lundmark, who had access to the Lick and Mount Wilson collections, noted the similarity between the irregular system NGC 4449 and the Magellanic Clouds. And the final proof came when Hubble, in his survey of galaxies, hit upon NGC 6822. This was the first system that he investigated for variable stars, and apparently it caught his fancy, for he wrote a number of small notes about it before he published the complete investigation. So finally the Magellanic Clouds sneaked in as extragalactic systems, although it took a long time before they were recognized as such.

Hubble had started his astronomical career with a thesis at the Yerkes Observatory, which was simply called "Investigations of Faint Nebulae." He took plates of one or two fields with the 24-inch reflector, and described the nebulae that he found in an elaborate and lengthy way. When he went to Mount Wilson he followed very much the same line, first with the 60-inch, later on with the 100-inch, exploring as many extragalactic nebulae as he could. He took a huge collection of plates at that time to get thoroughly acquainted with the galaxies. His first major result was the classification that we discussed in the last chapter. But he made no attempt to attack the problem of the galaxies directly.

The man who attacked the problem was Duncan, then at Wellesley; he used time in his vacations to search for more novae in the Andromeda Nebula. He found quite a number; it was a rich hunting ground. During this search, in about 1922, Duncan discovered a variable star of small amplitude, followed it up, and showed that it had all the earmarks of a Cepheid. That gave the final impetus to a direct attack on the galaxies.

Hubble knew that, besides the Andromeda Nebula, the

nearest systems on his list were the irregular galaxy NGC 6822 and the spiral M 33. Within four years he subjected all these systems to a first very careful survey and analysis, and found Cepheids in all of them. By finding all these Cepheids he established beyond doubt that the supposition that these objects were nonstellar was very definitely incorrect. The Cepheids showed normal amplitudes and it was clear that we were dealing with single stars. If we are dealing with a combination of two stars the amplitude reduces very rapidly, and a group of stars makes it disappear completely. Now we had the possibility of determining the distances for the first time.

In judging Hubble's survey of these three galaxies one must keep in mind that in the 1920's one could have photovisual plates, but their sensitivity was so low that exposure times were at least five or six times as great as those for blue plates that reached comparable magnitudes. The big advance in sensitized plates, on which we lean today so much, was only made in the 1930's. So Hubble's surveys were necessarily restricted to the photographic region; in each system he heroically took long-exposure photovisual plates, up to 4 hours, to get at least an idea about the colors.

After deriving the distance modulus of each system, Hubble went on to determine their other characteristics, especially with regard to the stars. For each system he tried to determine the upper part of the luminosity function, and found that in each case it was similar to that in our own Galaxy — the well-known van Rhijn–Kapteyn function. This identity of the brighter parts of the luminosity functions showed that the brightest stars in these galaxies should be good indicators of distance. He took into consideration the data which in the meantime had been obtained at Harvard, and concluded that the luminosity of the four or five brightest stars would be a very good indicator of the distance of the system.

With these brightest stars we run into some difficulties. If we restrict observations to the blue, we naturally include clusters of stars among the brightest objects. Even with the 200-inch, at the distance of the Andromeda Nebula we should be unable to resolve the Pleiades, which would appear like a little clump; with very fine seeing we would probably observe the brightest two stars, but with average seeing the cluster would be indistinguishable from a star. Also Hubble could not distinguish H II regions from stars. So his values for the brightest stars were not very reliable, although the principle was correct. He realized, and insisted, that if properly standardized with similar objects in our own galaxy M 31 would still appear reddened. Of course there is a problem in standardizing with a corresponding mixture in the Galaxy; we do not know what is the percentage of clusters, H II regions, and so on. But in principle the method is sound, and can be applied without difficulty.

Hubble also showed that these galaxies contain emission nebulosities and H II regions (of which he found quite a number in NGC 6822, and also in M 33), and he showed that star clusters occur in all these systems. Especially well known is his investigation of the globular clusters in M 31, in which he showed that their number is comparable to that found in our own Galaxy.

He showed, therefore, that spirals and Magellanic Cloud-type systems (the well-known types that he had already investigated) are clearly stellar systems of the same order of size as our own Galaxy. The problem of the nature of the spiral nebulae was settled. He also showed that the upper part of the stellar luminosity function is the same as in our own Galaxy, and that apparently the same constituents (Cepheids, H II regions, star clusters, both open and globular) make up spirals and our own system.

On one point he went astray, because in M 31 he was unable to find blue stars and H II regions (emission nebulosities). From this he concluded that there was at least a strong indication that O and B stars of high luminosity are present in *Sc* spirals and in the Magellanic Clouds, but are absent from the *Sb*'s. From his own investigations he concluded that the gas must be there, and absorbing matter was indicated everywhere in M 31 by the absorption bands. So he realized the presence of dust and gas; but since the ingredients of an H II region are gas and O and B stars, he concluded from the absence of H II regions that the O and B stars must be absent. Here he was a victim of the fact that he could at that time observe only in the blue; red plates of any sensitivity were not available at that time. His facts were perfectly correct, and only later observations showed that the reason lay in the absorption and reddening due to the presence of dust and gas, and in the high inclination of M 31, which exaggerated the effect.

Such is the qualitative side of the picture that emerged at that time; most of the results can be taken over even today. But there were disturbing facts on the quantitative side. Hubble had published a period-luminosity relation for the Cepheids in each of his systems. The relations for the Small and Large Magellanic Clouds had been obtained at Harvard. The disturbing thing was that all these curves had different slopes, so that it was impossible to compare two systems or determine their relative distances, because at some place the curves intersected, leading to quite a range of distances for a system. The general period-luminosity curve that Shapley had derived from his globular clusters again showed a different slope. It was obvious that the reason for these discrepancies was the uncertainty of the photometric scales that had been used.

In the first chapter I mentioned that Hubble was severely handicapped in his work on M 33 and NGC 6822 by the fact that when he started the only reliable sequence was the North Polar Sequence, which was good (as later photometric checks showed) only down to about $16^m.5$, whereas his Cepheids started around magnitude 18 or 19. And since he could not reach the Polar Sequence anyhow with the 100-inch telescope, he was forced to use the Mount Wilson catalogue of Selected Areas, and extrapolate the magnitude scale from nearby fields by hook or crook. The Selected Areas are of much lower accuracy than the Polar Sequence; more plates were taken at Mount Wilson for the Polar Sequence alone than for all the Selected Areas. By trying every possible method, Seares succeed in getting the scale of the Polar Sequence correct down to $16^m.5$; in the Selected Areas he had to restrict himself to four or six plates. If we plot the modern photoelectric magnitudes in the Selected Areas against the International System, we find that instead of being a straight line the curve is wavy; the period of the waves corresponds to the reduction by the diaphragms that were used for calibration. There were too few plates to overlap over the whole magnitude range. By extrapolating, and fitting either a hill or a valley in the wavy curve, one could end up with a systematic error of $0^m.3$ or $0^m.4$.

So it was not surprising to find differences in the form of the period-luminosity curve. Even today the question is not settled. Arp has made the first step in photometric work in the Magellanic Clouds; Sandage is trying to make the next. The problem is really very much more difficult than was at first realized. We thought that by merely using the photocell we should be safe, but the unresolved background, which varies in intensity from point to point, has caused tremendous trouble. We may hope to eliminate the trouble to a large extent if the Cepheids are very carefully selected. In the next few years we hope to obtain the mean period-luminosity curve for which we have looked so long.

I think everything indicates that there *is* a mean period-luminosity curve. There are some things in the literature about large differences, such as Kukarkin's results, which come from using very poor material — individual data from Hubble. Statistically, Hubble's results are all right, but one cannot use individual light curves. The investigations of Cepheids with the 200-inch now show very close similarity between the Andromeda Nebula, the Magellanic Clouds, and our own Galaxy. There may be a slight shift of about half a day in some of the relations, but I am not sure whether it is real or imagined.

The nonparallelism or nonidentity of the forms of the period-luminosity relation can easily be brushed aside as an imperfection of the time. But soon other things showed up which indicated that the trouble went much deeper than the matter of magnitude scales; somewhere along the line of reasoning there had been a fundamental slip. The most striking was the discrepancy between the luminosities of the globular clusters in the Andromeda Nebula and in our Galaxy.

In our own Galaxy we can determine the luminosity of a globular cluster even if it is heavily obscured. On the same photometric system we determine the total magnitude of the cluster and the brightness of the cluster-type variables. The absorption cancels out, and we obtain the total absolute magnitude without any difficulty. Hubble found in M 31 that the globular clusters extend far beyond the disk, so there is an area where we can determine magnitudes, free of any interference. It turned out that when we use the distance obtained from the Cepheids there is a difference of about 1^m.5 in the upper limit of the luminosities of the globular clusters in M 31 and in our own Galaxy, in the sense that the clusters in M 31 would be 1^m.5 fainter. It is usually stated that the globular clusters as a whole would be 1^m.5 fainter. Let us be very strict on this question; nobody knows, even today, the total number of globular clusters in M 31, or the lower limit of magnitude; we do not even know it in our Galaxy. But the difference is very well established for the upper limit, and in this case that is sufficient, because we have first-rate evidence that the two samples are comparable in numbers. The upper limits differed by 1^m.5, and that was very disturbing. It was very hard to see how one group could differ systematically in the upper limit from the other, because Hubble had made the first measures of the colors of the globular clusters, and they were similar to those for our Galaxy. Humason had taken the first spectra, and the spectra agreed too. So there was an outstanding difference.

The same difference was indicated by the novae in M 31, of which by that time 100 had been discovered, but the data were not quite as clear. The novae in M 31 seemed to be at least 1^m fainter absolutely at maximum than those in the Galaxy, and the luminosity of those in the Galaxy was by that time fairly well established.

Finally there was the unusual size of our own Galaxy, very much larger than anything that was obvious in the sky, and this had plagued us for a long time. In the mean-

time the red-shift–magnitude relation had been standardized on the same system, and again our Galaxy showed an
intolerable deviation.

Such was the situation at the end of this first phase. Qual-
itatively everything in M 31 agreed with our own Galaxy,
but the quantitative measures were in very serious difficulties. By that time Hubble had left this field, which he would
undoubtedly have cultivated more if he had not in the
meantime become absorbed by the red shift and the expansion of the universe, and he now turned entirely to the
cosmological problem.

Now these discrepancies interested me exceedingly at
that time, and since I wanted to stay out of Hubble's field,
cosmology (in which I was anyhow not interested), I tried
to see how one could at least clear up some of these discrepancies. One of the first attempts that I made was to see
whether I could get better magnitude scales.

I had many discussions with Seares, but the methods
that he had used to establish the scales down to $16^m.5$ or 17^m
(putting either a huge screen or diaphragms over the telescope) were not usable. As one tried to extend the scale to
fainter magnitudes one had to take long exposures, the first
on clear film, the second on film that had been preexposed.
This preexposure and postexposure create an insurmount-
able problem. The recent paper by Elvius shows how
serious the problem is; he tried to get a better zero point
for the scales of different selected areas by transferring from
the pole directly to the plate. In spite of all precautions,
these preexposure-postexposure effects were as great as $0^m.5$
— it is unbelievable how large the error can be.

The only method which would have worked would be
one in which the exposures were simultaneous, say with a
grating in front of the telescope. But this is very wasteful

of light. Then I met Heckmann, who had had great success with the platinum half-filter. He showed me the unpublished paper of Heckmann and Haffner on the color-magnitude diagram of Praesepe, work of very high precision. Hopefully I started establishing scales with the platinum half-filter, but soon found that it solved my problem only to a certain degree; I still had the preexposure-postexposure problem. Heckmann was free of it, because he used a refractor of focal ratio 12, and very slow plates of very fine grain; he stayed in the brighter range and so he always got away from sky fog. But with an $f/5$ reflector, and trying to push to magnitudes 20 and 21, I hit it.

Fortunately the method can be modified in such a way that it "rings a bell" if something goes wrong. You make an exposure through a half-platinized glass plate, then turn the whole thing around and make another exposure. When you do it, you see see that the two sides have a huge difference in the sky fog, so just as a tranquilizer you turn the filter around (you have the whole preexposure-postexposure effect again) and go on. What you need in addition to this plate is a second plate, taken without a filter; you determine the absorption in the glass and derive the scale. Unfortunately you derive this independently for each half of the plate, and get two scale determinations for your field. Generally the two agree completely up to a certain point, and then they diverge; you are always safe up to that point. It was quite clear that the preexposure-postexposure effect was involved, because the lower the sky fog, the farther this point was — the later the breakdown.

Finally with much pain I had a system of magnitudes reliable to about $20^m.2$ for Mount Wilson Standard Area 68, conveniently located for M 31 and most of the members of the Local Group. It cut off very severely here. It was now only a question of getting a decent zero point. Finally in 1938 Stebbins, who was always willing to help, determined a star of 16^m.1 in Standard Area 68; it took a whole night with the 100-inch; this was before the advent of the 1P21. I remember that Seares, Stebbins, and Whitford told me not to trust it too much; there might be an error of 0^m.2. But I considered it strictly as zero because it was better than anything we had ever had. So I had a scale which might decrease some of the major discrepancies that could have crept into Hubble's work on M 31. By repeating the transfers to M 31 with the new scale I was able to get an idea of the corrections to Hubble's scale in the most important part of his work. The corrections were quite large, up to 0^m.7 or 0^m.8 at 20^m. But today we know that this was still an underestimate, because I relied too much on this one photoelectric magnitude, and we actually had to add 0^m.2 more.

Simply by using the new scales, and the brighter part of the period-luminosity curve, the distance modulus had gone up, I think, to 22^m.7 instead of Hubble's 22^m.2. I did not run the fainter end, where *my* scale again was wobbly. Obviously these scale corrections were important, but they were not the real source of the discrepancies. The cause of the difficulties lay much deeper.

In the course of this work on the members of the Local Group, one remarkable thing had shown up: in the Andromeda Nebula, Hubble had found it impossible to resolve the central part. Also — even more significant — he had been unable, in spite of many attempts, to resolve the companions, which were the closest representatives, and apparently normal representatives, of the *E* galaxies. It was quite clear that here was a problem that had to be solved, and a place where one might expect to find an answer to the other problems that he had encountered. Why could he not resolve the inner part of M 31? I had innumerable discussions with friends. We were all bound still by the ordinary Russell diagram, and the verdict always was, "Oh, well, somewhere at fainter magnitudes the thing must begin." A special stumbling block, which always deterred a real attempt, was the classification of the large-dispersion spectra of M 31 by Hubble, Adams, and Joy as dG5. That was correct. One could expect to find only the brighter stars, and to reach the main sequence around G5 at $+5^M$ was completely hopeless. This spectral classification really held up a serious attempt for a long time, because we were all convinced that if the brightest stars were dG5 we could not go sufficiently far down the sequence.

CHAPTER 4

PHOTOGRAPHY OF GALAXIES

By 1936 it was clear that part of the inconsistencies that had shown up in Hubble's pioneer work were simply due to the use of provisional photometric scales. But for the rest, the causes really lay very much deeper, and to the best of my knowledge nobody at that time even ventured a guess as to what the trouble really was.

In order at least to settle the question of the scales, I decided to establish new scales by the platinum half-filter method, and, in cooperation with Harold Weaver, I spent some time after 1938 in establishing new photometric scales in a few strategic Selected Areas (S.A. 68 to connect M 31 and M 33 and IC 1613, and especially for the winter sky, S.A. 57). Even the platinum half-filter method did not work too well with a reflector, especially for faint stars, because of the preexposure-postexposure effect, but at least it was possible in this way to get reliable scales down to about 20^m.2 for the first time. The uncertainty was in the zero point.

By 1942 the sequence for S.A. 68 was available, and I proceeded to check Hubble's scale in M 31; I used the 100-inch telescope, diaphragmed down to 84 inches, to get a coma-free field in the center. To avoid the preexposure-postexposure effect one half of the plate was exposed to the Andromeda Nebula for 90 or 70 minutes, and then the plate was turned around, the other half was brought onto the optical axis, and an exposure of the same duration was made on S.A. 68. Ordinary 103a-O plates were used. The resulting corrections to Hubble's scale are given in volume 100 of the *Astrophysical Journal*.

Now these plates were taken under really excellent seeing, and in examining one of them — a 90-minute exposure of M 31 — I was very much surprised to find a big piece of spiral arm in the region north of the area in which Hubble found most of his variables. On a plate taken with the 200-inch an area of amorphous nebulosity is seen adjacent to the resolved spiral arm; in this area (to my surprise) there were for the first time signs of resolution into stars on the 100-inch plate. What is meant by incipient resolution can best be understood from the reaction of everybody who looks at this photograph. The plate is very irritating to the eye; a definite structure is emerging all over, but one does not yet see any stars. After the area had been resolved, the reason was perfectly clear: in any field of star distribution, even the smoothest, there are of course fluctuations of brightness and density. And these areas of higher density, areas where there are several bright stars, are the first to come out as incipient resolution; what one sees first is this indefinite pattern.

I can best illustrate by giving a specific example which was very illuminating. In the round companion of the Andromeda Nebula, M 32, a kind of curved diffuse shape has long been known, which was weakly visible even under average seeing conditions. On a plate taken under very much better seeing, this feature, normally very soft and diffuse, narrowed up and became much sharper, but was not really different. When resolution was finally achieved, the feature turned out to be one of those accidental chains of stars of nearly equal brightness, with some fainter ones in between. This hazy filament showed up before resolution was finally achieved.

The resolution of M 31 was now clearly within our grasp; the question was, how could it be attained? It was clear that no further advance could be expected with blue-sensitive plates: under wartime conditions at Mount Wilson, when the Los Angeles valley was completely blacked out, a 90-minute exposure was about the limit set by the sky. Any further increase of exposure time would not have been warranted by the gain in magnitude. The only hope was to try something else, and I decided to see whether red plates would do the trick, which would work, of course, only if the brighter stars to be resolved were actually red stars. The fastest plate available at the time was the Eastman 103-E (not yet the 103a-E), which had the advantage of being very stable, and could be stored in a cold room for years, because it was not fully ripe and up to full speed. It had to be ammoniated, and with proper ammoniation it could be speeded up from 1^m to $1^m.5$; I always stepped it up to optimum performance by ammoniation. Even so, at first sight it was still considerably behind the blue-sensitive plates.

Imagine that you had taken a 1-hour exposure of the

Andromeda Nebula in the blue, and a 1-hour exposure in the red (103-E + RG2). If m_{pg} is the limiting magnitude with the blue 103a-O plate, then the limiting magnitude achieved in the same exposure time with the 103-E + RG2 turned out to be $m_{pg} - 1^m.6$ for color index zero (an Ao star). So the red plate was $1^m.6$ behind for an Ao star, which did not look very hopeful. But whereas the sky fog set a limit of 90 minutes to the exposure time for the 103a-O plates, the limit for the red plates was of the order of 8 to 9 hours for the same sky fog; one could gain at least a magnitude by increasing the exposure time. For the 103-E plates, to stay within a reasonable limit, I used about 4 hours, and it turned out that I could gain about $1^m.3$ by virtue of the exposure time.

For the relation between photographic and photo-red magnitudes we had the relation: $m_{pg} = m_{pr} + C.I.$, and for m_{pr} we may take $m_{pg} - 1^m.6$. By increasing the exposure time it was possible to reduce the factor $1^m.6$ to about $0^m.4$: the red plate was only $0^m.4$ behind the blue-sensitive plate, even for Ao stars. Thus the plates would be equal for a star of C.I. $0^m.4$, and for stars of larger color index I could reach fainter magnitudes. If the stars were really red, say of C.I. $+ 1^m.5$, I could hope to gain something of the order of 1^m over the stars reached in the blue. That was the idea on which I worked; the only chance I had was to detect stars of large color index on red-sensitive plates.

By that time it was known already that the color index of the inner region of M 31 was of the order of $+0^m.9$, and there was every reason to suppose that most of the light came from the brighter stars; this would be so with any likely luminosity function. Although this argument was by no means conclusive, one could expect the color index of the brightest stars to be at least $+0^m.8$ or $+0^m.9$, which would have given a gain of about half a magnitude over the blue plates. But however one made the guesses, it was still touch and go.

No success could be expected by simply relying on these guesses, snapping a red plate into the plateholder of the 100-inch, making the exposure, developing, and hoping to see something. It was quite clear that the stars would be very faint, and also very likely that they would be extremely crowded. It would tax the resolution of the 100-inch, and evidently one would have to be very careful and use every trick of the game. In order to keep the resolution as high as possible it was necessary, first, to make observations only under the best seeing conditions, when the confusion disk of the stars was very small. Second, observations could be made only on nights when the figure of the mirror was near to its ideal state, with no upturned edges — something which always increases the size of the image. Third (and this was the main problem), something had to be done about the changes of focus, which exist because the 100-inch has a mirror made of old-fashioned glass. Even on good nights they run from 1.5 mm to 2 mm during the night, and there are nights when they run up to 5 mm or 6 mm.

Everything was determined by the resolution of the plates. For the 103a-E plates, the smallest images that can be obtained lie between 30 μ and 40 μ. Eastman Laboratories determine the resolution by putting a sharp edge against a photographic plate, exposing, and finding by photometry how sharp the image is. This gives the resolution for a linear edge; for a circular image one doubles the value and arrives at the figures given above. The scale of the 100-inch at the Newtonian focus is 1 mm = 16″, so the smallest images as determined by the resolution of the plate would be of the order of 0″.48 to 0″.64, depending on whether the resolution was 30 μ or 40 μ; of course there was no sense in trying to push them further. This meant that the confusion disk of the stars during observation had to be of the order of 0″.5 to 0″.6 — say smaller than 0″.6 — over the whole 4 hours if possible. Fortunately we had quite a number of nights of this quality during the late summer and fall months, although they were not frequent; the confusion disk of the stars was smaller than 0″.6 either during the major part of the night or sometimes for the whole night.

Now let us consider the focus. The 100-inch has an $f/5$ mirror, and under good seeing conditions it is easy to focus down to 0.1 mm or even a little better; you use a knife-edge and make a run, and the focus will fall within 0.1 mm. Since you are working with an $f/5$ mirror, the confusion disk resulting from error in focus is of course one-fifth of 0.1 mm; so the maximum confusion disk from imperfect focus is 0.02 mm. If you can hold the focus to 0.1 mm you are safe, because the resolution of the plate is 0.03 to 0.04 mm.

The problem of focusing was rather a serious one. First I tried the scheme that Ritchey proposed, and used extensively, to make the beautiful pictures that have been reproduced in all the textbooks. He had a special plateholder made which could be swung out and put back into exactly the same position, so that he could swing it out during the exposure when he had the feeling that the focus had run off the knife-edge, set the focus anew, swing back, and continue. But I found that the method did not recommend itself for my purpose; the whole manipulation had to be performed in the dark for the red plates — going to the new field, making the settings, making the readings — and by the time you have done all this you have lost about half an hour, and when you start you are already off the best focus again.

So I fell back on the old method used by Keeler in the early days of the Lick Observatory; I followed the focus changes by watching the coma of the guide star that I used. The comatic image can have various forms, according to how the focus is set. The comatic image is bordered by two rather intense lines caused by diffraction, and inside there is a second set. I can confirm what Keeler stated, that the position of the two bordering lines is very critical for changes of focus. If the focus changes in one way or the other, the inner line begins to cross the outer, or recedes far inside it. So the trick was to set in such a way that the outer line was asymptotically tangent to the inner line. After I had focused with the knife-edge, I immediately set the focus of this guide star so that the inner diffraction patterns were tangent to each other asymptotically near the center.

I should like to explain why this method has fallen into disrepute. In order to see these lines in a critical way, you must have very good seeing, and you must use very high magnification. In guiding I always used a magnification of 2800 in order to see these lines, and the slightest disturbance prevents one from seeing them. My procedure was to make sessions of 4^h to learn the game of following the focus, sometimes on mediocre nights; I started with the knife-edge and then took over with this method, just changing the focus all the time as the image demanded.

I remember the first night very well; I selected intentionally a night when the seeing would presumably not be first-rate. I still remember how confused I was as to what to do. But you just calm down and wait for a moment of good seeing; and after you have mastered it, it is astonishing how little time you need to see what the situation is, and to give a quick turn to the focus. By training myself in this way I finally learned to handle the method. Under ideal seeing conditions I am sure that anyone could do it after a few sessions. The whole trick is in dealing with the occasional disturbances; I can imagine that some people would lose their heads and get nervous and angry in a short time, and would begin to turn the focus violently.

I was very much pleased after one of these exercise sessions to find that I had gone on for 4 hours by simply manipulating the focusing screw. It turned out afterwards that the focus had varied by nearly 6 mm that night; at the end of the session I read my focusing screw, put in the knife-edge right away, and found that it was still within 1 mm. So this method has very much to recommend it; it is quite easy when once you have mastered it. You must start right away with very large magnification, which means that you cannot use excessively faint guide stars; because of the slight deterioration in the seeing, you would no longer see the details in a faint guide star.

I had a special plateholder made — not anything fancy, but exceedingly precisely machined — and the corresponding knife-edge very carefully adjusted so that the focus determined by the knife-edge corresponded to the film of the plate, also within these limits.

The other problem was to keep the proper figure of the mirror. One of the big troubles with large reflectors in domes that have no temperature insulation is the daily rise and fall of temperature, which of course affects the mirrors. The main trouble for all Newtonian arrangements is in the secondary mirror, not so much the main mirror which is farther down, and does not respond rapidly to changes of temperature unless the changes are very fast. The large focus changes I have mentioned are immediately due to the daily heating of the secondary mirror, which is in a rather high position. In the morning hours the dome heats up so much that this mirror goes from its plane figure into a hyperbolic figure (or whatever it is); it then returns slowly, producing changes in focus and astigmatism.

I have already mentioned that we have nights of fine seeing at Mount Wilson often enough, and one can fairly estimate when such a period is coming. The cycle in Southern California during the summertime runs something like this: there is a period of slow warming up, and you get fine seeing near the top of this warming-up period. It may stay on the top for 3 or 4 days, sometimes a whole week; then very suddenly comes a cooling-off period, when the temperature falls 10° or more in a day. During this cooling-off period you are lost, you have a mixture of different air masses, and the seeing gets poor. But in general, if you are on the mountain for some time, you can pretty well tell when such a period is approaching. For this period we had to keep the figure of the mirror in as good shape as possible. Since we have no insulation, the only thing to do was to open the dome early — right after lunchtime — and turn it around so that no sunlight could fall directly on the instrument. We have of course the advantage of a mountain situation: the moment you turn into the shade the air is cool; there is a big difference between the air temperature and the temperature in the sun, and the air temperature does not change much between day and night. So the secondary mirror had time to cool off already in the early afternoon. Usually in the time interval between about 1 p.m. and 10 p.m. (when the instrument was to be used) the secondary mirror had a chance to get into equilibrium again. These preparations lasted from the fall of 1942 to the fall of 1943, when I was finally ready to make the real test. Indeed, after taking all these precautions, the resolution was a very simple matter. In August and September I resolved the central part of M 31 and the two nearby companions in rapid succession. Mayall had told me that NGC 185 (which forms a pair with NGC 147, some 12° from M 31) has a radial velocity very similar to that of M 31, and that he suspected on this basis that they might be nearby and associated with the Andromeda Nebula system. So I tried NGC 185 and NGC 147, and indeed they were easy to resolve.

After the shooting was over, it was quite clear that all the precautions had actually been necessary; I had just managed to get under the wire, with nothing to spare. It shows that there are always situations where you simply have to have good techniques.

By the end of the first season it was clear that actual resolution had been achieved. I had taken two or three plates of NGC 205, and on these plates I found the first half dozen variables. This was simply a sign that these objects were actually single stars, but in many cases we were still observing clumps of stars. Subsequent work with the 200-inch has convinced everybody that there is no doubt as to the resolution.

It was a remarkable thing that, as soon as resolution was attained, stars appeared in very large numbers — thousands and tens of thousands. This obviously meant that we were reaching a region of very high density where there were red-giant stars, otherwise it was inexplicable. There is one beautiful concentration in the H-R diagram: on the red side are the normal giants, but if you go higher up to the supergiants there are no concentrations of stars. It was some time before I realized what these red stars could be. They were certainly not ordinary giants, which would have been too faint to be reached at all. The whole question was: what else could they be?

I had always been very much interested in the globular clusters, but in those days I had almost forgotten them. After a long time it finally dawned on me that there is another concentration in the H-R plane, what we now call the giant branch of the globular clusters. The moment I thought of that, everything began to make sense, because the red giants in globular clusters also agreed with the observed brightness. Normal red giants of absolute magnitude o were completely out of the question; the distance was too great, even allowing for uncertainties in the magnitude scale. I do not know what would have happened if we could not have ruled them out; I am certain that both I and everybody else would have stuck to identifying the stars with ordinary red giants. But it was out of the question; and then I remembered the globular clusters, with their concentration of red giants at -3^M, just about in the range where we should expect these stars to be. Because of the asymptotic approach of the red-giant branch to the upper part of the diagram, the moment you cut into it you immediately reach large numbers of stars.

This was of course not sufficient to clinch the case, so the next step was an attempt to reach another part of the color-magnitude diagram of the globular clusters, which proved to be a very simple thing. Shapley, in his early work on globular clusters, had worked out all the important relations between the constituents; he knew that the cluster-type variables were in the mean about $1^m.5$ fainter photographically than the brightest giants; the photovisual difference would be $3^m.0$ because the color index of the red giants was $+1^m.5$. Was there any chance of reaching the cluster-type variables? Because we had only reached the red giants with great difficulty, the cluster-type variables were unattainable.

The Sculptor and Fornax systems had been discovered at Harvard a few years back. Their nature, which was not clear at the time, has become so through the analysis of NGC 185 and NGC 147. The two latter provide the links between the E galaxies of intermediate luminosity like the bright companions of M 31 and the E galaxies of very low luminosity, which are represented by the Sculptor and Fornax systems. It became clear that we should have to classify these, and the similar ones found later, as dwarf E galaxies. Here the situation was really ideal, for by that time Hubble and I had found cluster-type variables in the Sculptor system, and they lay photographically $1^m.5$ or $1^m.6$ below the giants. So, granting that the Sculptor system was simply an E galaxy of very low luminosity (and there was no longer any doubt about it), we had a chain of evidence that the stars that had been found in resolving M 31 were similar to those in globular clusters. First, the upper limit of magnitude was of the right order; second, such systems also contained cluster-type variables whose magnitudes bore the proper relation to those of the brighter stars. So you might put it that the color-magnitude diagram was fixed by two points: the giants and the cluster-type variables, especially the latter.

The ridiculous part of the situation is that, after all these efforts, it appears that the whole difficulty should have been unnecessary When Shapley discovered the Sculptor and

Fornax systems he described them as a new type of stellar system because they looked so peculiar. Hubble and I were of course immediately interested, set out to make a rapid survey, and discovered the cluster-type variables. We treated the Sculptor system as if it were a normal globular cluster; we found the cluster-type variables, noticed that the brightest giants were 1^m.5 to 1^m.6 brighter, and never realized what we had got hold of. We described the systems, even though they were in some ways different, as exhibiting the well-known features of globular clusters. The paper was soon published and was discussed by everybody, but nobody had the wits at the time to see that the whole problem was solved and that we knew what the E galaxies consisted of. This strenuous detour from another angle was needed before we saw the light. I think this should be a consolation to the student who sometimes feels a little bit discouraged because things are not going so well. As Einstein said, mankind is very stupid and progress is very slow.

You know what finally happened. After putting together the things that we otherwise know about these galaxies, I concluded that we had to distinguish at least two H-R diagrams — one the normal diagram that we had known well for some time, the other, the globular-cluster diagram. The further inference at that time was that the E galaxies were essentially made up of stars of globular-cluster type (Population II), and that in the Sb and Sa spirals we had the two populations together, distinguished as groups by their spatial arrangement, with Population II essentially predominating in the central region. The remarkable fact was that some systems contained only Population II and in other systems the populations were mixed. At the time we could not guess what the situation would be in systems like the Magellanic Clouds — among the most splendid Population I systems that we know of; this population makes such a splash that it would undoubtedly cover the very much fainter Population II unless one searched specially for the latter. I guessed at that time that Population II would also be present in *Sc* spiral and Magellanic Cloud types, but covered up by the supergiants of Population I. It was a first rough picture; a very interesting feature was the existence of a whole class of galaxies which (as I cautiously expressed it) had a color-magnitude diagram closely related to, if not identical with, that of the globular clusters.

It became evident that in this division we had found something very much deeper, when we considered the relation of these populations to the presence of dust and gas. Hubble's investigation had already shown statistically that the *E* galaxies are essentially free of dust and gas; on the other hand there was plenty of evidence of dust and gas in the spiral arms, because dark obscuring patches were visible everywhere, and it appeared that Population I might be associated in some degree with this dust and gas. So the most intriguing result was the relation between the two populations and the dust and gas: Population II was characterized by absence of dust and gas, Population I not yet quite certainly by its presence. In the next chapter we shall see how a special investigation of M 31 actually showed that there is an intimate connection between spiral structure, dust and gas, and Population I. The special survey of the Andromeda Nebula then led rapidly to the conclusion that the primary thing is the dust and gas; the stars are a secondary phenomenon. In this way a clearer picture emerged of the relation between dust and gas and the two stellar populations.

CHAPTER 5

THE ANDROMEDA NEBULA:

SPIRAL STRUCTURE

THE two stellar populations are defined by their H-R diagrams; in this sense the definition is a physical one. We knew also that we can divide star clusters (open and globular) into two groups on the basis of their color-magnitude diagrams. This is very satisfactory; the two kinds of star clusters, which have been distinguished for a long time, appear as representatives of two large stellar populations.

The remarkable relation between population and the presence of dust and gas definitely pointed to something still deeper. The *E* galaxies, which were thought to consist of pure Population II, are essentially free from dust and gas; when they contain a little, as indicated by bright λ 3727, the amount is negligibly small compared to the mass contained in stars.

In order to get a deeper insight into this relation between dust, gas, and stellar populations in an individual galaxy, I decided after doing the work on resolution to make a new survey of the Andromeda Nebula at the 100-inch. I will mention two of several reasons for making such a survey. In the first place it had become clear that M 31 extended much farther in all directions than appears in the usual pictures. A second reason was that, to my great surprise, the first plate that resolved M 31 showed three or four big emission nebulosities which I had never seen before. I have mentioned that Hubble, who could work only with blue-sensitive plates, had made a very careful survey for emission nebulae in the so-called main body of M 31, and had been unable to find a single one. Since the well-known ingredients of an emission nebula are the presence of O and B stars, dust, and gas, he concluded that O and B stars are not present in Sb spirals, because there is plenty of evidence of dust and gas. Obviously there was now a chance to make a real search for emission nebulosities.

My guide to the actual extent of M 31 at that time was a remarkable paper published by Vocca in Bologna. He had used a 4-inch telescope and had photographed M 31 with exposure times just reaching up to the sky fog. Stage by stage he stepped up the negatives, working with great care to deduce all the information, and really did a first-rate job. When I measured up his figures I found that probably the outermost isophote that could be seen on these stepped-up prints of M 31 was a very elongated affair, 280' × 56'. This picture really guided me very much in laying out my plan.

My plates, made with the 200-inch, were oriented along the main axis of M 31. The dimensions of each plate covered 45' (along the major axis) × 32' (parallel to the minor axis), with sufficient overlap between adjacent plates.

This required essentially ten plates on either side of the major axis (from north-following to south-preceding) if I wanted to cover the area of M 31.

There actually were four series of plates: (1) ordinary blue-sensitive exposures, each 60 minutes, 103a-O + GG1 (the GG1 was simply used to stay on the photographic system because of the aluminized mirror); (2) a series of 4-hour plates, 103-E + RG2, covering $\lambda\lambda$ 6400–6700 and including Hα; (3) another 4-hour series, 103-U with a combination of two filters, covering $\lambda\lambda$ 6700–7200; (4) ultraviolet exposures of 2 to 3 hours, 103a-O + IG2, $\lambda < 3900$, covering interesting regions only. It was quite an ambitious program. I added another series on either side, parallel to the major axis, but only in the blue, to search for outlying globular clusters. In addition to this I added another series on either side of and parallel to the major axis, farther out, and a little bit narrower.

Series (2) was essential for reaching Hα. To pick out the emission nebulosities I used a region free from any strong or intermediate emission lines; one of these regions is $\lambda\lambda$ 6700–7200 (series 3), exceedingly favorable because it is very narrow, and contains only very weak nebular lines. So comparison of 103-E plates with 103-U plates could show emission objects strong on the 103-E, and very weak or absent on the 103-U.

It was to be expected, of course, because of the large distance of M 31, that a considerable percentage of the nebulosities would be very small — on the plates they would appear semistellar or even stellar.

Of course, the requirements of seeing were not so stringent for this problem as for the resolution, which I described in Chapter 4, so it was possible to use good nights; I only rejected mediocre and poor nights. I started with the 4-hour exposures in 1945, and just finished in the fall of 1949, when the 200-inch came into operation.

I will first give a general and schematic description of the spiral arms of M 31, as they showed up in this extended survey. I shall name the arms in order as they cross the major axis: on the north-following side I call them in succession N_1, N_2, . . . ; on the south-preceding side, S_1, S_2, . . . I may say at once that from N_2, S_2 onward these are crossings of the same arm, but it is doubtful whether N_1 and S_1 belong together. Otherwise the pairs with similar indices belong together. Whether the higher ones, like S_6 and S_7, are returns of one of the earlier ones, we do not know at present. We must await the complete disentanglement of the spiral structure.

Table 2 gives the approximate distance of each spiral

Table 2. Distances of spiral arms
from the center of M 31.

Arm	Distance (min)	Distance (kpc)	Arm	Distance (min)	Distance (kpc)
N_1	3.4	0.6	S_1	1.7	0.3
N_2	8.0	1.5	S_2	10.5	1.9
N_3	25	4.6	S_3	30	5.5
N_4	50	9.2	S_4	47	8.6
N_5	70	12.9	S_5	66	12.1
N_6	91	16.7	S_6	95	17.4
N_7	110	20.2	S_7	116	21.3

arm from the center of M 31, in minutes of arc and kiloparsecs. For the linear distances I assume for the distance modulus $m - M = 24^m.0$, corrected for absorption inside M 31 and on the way to our Galaxy.

Now I give descriptions of the arms, which we can take in groups as indicated. I shall just give a general description of what one can observe now about the appearance of these arms.

N_1 and S_1: Dust arms; no supergiants or H II regions visible. (Here I mean by supergiants not only what is usually meant, but also bright O and B stars — in fact, the whole complex of exceedingly bright stars.)

N_2, N_3 and S_2, S_3: Dust arms with a sprinkling of supergiants. The first supergiants appear in N_2, S_2; the first H II regions appear in N_3, S_3. These are still essentially dust arms, with a sprinkling of the things that mean that star formation is going on.

N_4, N_5 and S_4, S_5 (intermediate arms): Maximum display of Population I; spiral arms now represented by abundance of supergiants; dust becomes more and more inconspicuous.

N_6, N_7 and S_6, S_7: Spiral arms defined by scattered groupings of supergiants, many of them blue; no obvious signs of dust (absorption patches) are visible.

You see for one thing that the arms extend very far out. Altogether you can distinguish seven arms running out from each side to about 20 or 21 kpc. The description reflects very beautifully the dual nature of the spiral arms: on the one hand, as congregations of supergiants, stars of Population I; on the other hand, as congregations of dust and gas. Clearly, as we go toward the outside, the sequence is: first, essentially a prevalence of dust and gas with only a sprinkling of supergiants, stars of Population I; then we pass to the maximum display of Population I, with the dust and gas beginning to fade out; and finally the outermost arms, where both Population I and the dust and gas are apparently fading out.

I ran the spiral arms only out to a place where I knew very clearly that there still *was* a spiral arm. Obviously the spiral structure extends still farther out, but so faintly that I could not pick it up with the present means. As evidence for this I found a cluster of B stars outside S_7 — not very bright, probably -4^M. This little cluster, which we should probably call an association, about 200 pc in diameter, is $2°.23$ from the center, corresponding to a distance of 24.6 kpc. There is also a beautiful string of B stars very far out in the spiral structure. It defines a little strip, also not very bright — the stars are -3^M or -4^M — which I am sure is an old "fossil" arm of M 31, long decayed.

The largest distance along the major axis to which I traced the globular clusters is a place $2°.5$ from the center, which corresponds to a distance of 27.5 kpc, where there are two globular clusters. They are very close to the edge of the system.

Table 2 shows a departure from regularity: S_2 and S_3 are farther out than N_2 and N_3; N_4 and N_5 are farther out than S_4 and S_5; S_6 and S_7 are again farther out than N_6 and N_7. Behind this departure lies an interesting disturbance in the spiral structure on the south-preceding side, which comes out very clearly when one tries to determine the position of the major axis from the turning points of the various arms. When the center is used as the fixed point, it turns out that all the turning points lie neatly on a straight line, except S_2, S_3, and S_4, which are displaced to one side; S_5 lies again on the straight line. The spherical companion of the Andromeda Nebula, M 32, is near the turning point of S_3, and it is tempting to think that it exerts a tidal action on these spiral arms. When I showed it to Schwarzschild, he made an estimate of the mass of M 32 from these tidal disturbances.

The final distances for the arms will have to await the analysis of the spiral structure. One can define the arms by the dust and gas, or by the blue stars. I cannot yet answer the question of how to take the final mean. The essential feature is the point at which a group crosses the major axis, and there are uncertainties for the very faint groups of B stars; if there were a denser group on one side of the axis, it would undoubtedly have shifted the apparent center over. What we have now is a schematic general picture.

I have already mentioned the dual nature of the spiral arms, which are partly dust and partly Population I stars. A most striking example can be seen by looking very carefully at the behavior of N_4, S_4; there is a bright outline on the top, with NGC 205 on the right. The top side is the receding edge of M 31. The arm N_4, S_4, as it swings round, is defined up to a certain point by the bright stars of Population I. Then suddenly it fades out and continues as a dark lane, with an occasional sprinkling of supergiants, and then (at about the bright star cloud NGC 206) it resumes its old form and is outlined by stars. It is exceedingly instructive, and we shall see later that this dark lane, which is a real spiral arm, is densely filled with emission nebulosities. So the arm really looks like an absorbing arm; it is really impressive.

I mentioned earlier that Hubble was not able to find any emission nebulosities, but my survey has yielded close to 700 of them. Today we know where these nebulosities are from the red plates, but if you look in their positions on the blue plates of the central body, where Hubble mainly searched, you can at best discover a very faint smudge that you would never recognize independently as an emission nebulosity. No wonder that Hubble missed them.

As the search proceeded, the emission nebulosities

showed up as most beautiful tracers of spiral arms; they are strung out like pearls along the arms. In the dark lanes associated with N_6 and N_7, where the emission nebulosities are, the dust establishes the position of the spiral arm. In this region, Hubble confused the intermediate area with a spiral arm. The structure is really not obvious; the spiral arm is like a chameleon — it can change its appearance. The reason must lie partly in the structure of the arm; one explanation would be that the dust curls over at the edges, which may be true and needs special investigation, though it sounds like an *ad hoc* explanation. But inside the arm you can see that the number of H II regions is just as great, in whichever direction you go along the arm. So the number of supergiants must be the same all along the arm, but the dust can change the appearance very greatly.

Emission nebulosities are really some of the best tracers of spiral structure; Morgan, Osterbrock, and Sharpless used them successfully for the first time in our own Galaxy.

The two most distant emission nebulosities found in this survey of M 31 were in the outermost arms, N_7 and S_7. If we denote by X the angular distance along the major axis, by Y the angular distance along the minor axis (positive in the direction of M 32), and by R the distance from the center, their positions are:

In N_7: $X = +116'.76$, $Y = +6'.35$, $R = 21.5$ kpc;
In S_7: $X = -113'.64$, $Y = -0'.52$, $R = 20.8$ kpc.

The extent inferred for M 31 is not of course final, but it is of the same order as was found from groups of stars and from globular clusters.

It is remarkable that these things can be observed in the red and not in the blue, and it is clear now that the reason is that they are simply reddened. This was proved when Mayall, who wanted to use the emission nebulosities to determine the rotation of M 31, found that their spectra were the same as those of the usual emission nebulosities of our own Galaxy. The reddening is simply caused by absorption, and the explanation is the more acceptable because of the high inclination of M 31. The angle between the line of sight and the central plane is only 11°.7, so we are looking at a flat sheet, and all the absorption effects are enhanced.

The sequence in the spiral structure begins essentially with dust in the inner arm, then stars of Population I become prominent and dust grows less conspicuous, and finally both fade out. This sequence also shows up very nicely in the study of the emission nebulosities, and emphasizes that the difficulty of photographing the nebulosities in the blue is caused by reddening. In the outer arm of M 31 the emission nebulosities are just as visible on the blue plates as on the red, though of course their numbers go down very much. In the innermost arms there are very few emission nebulosities; in the intermediate arms, with their maximum display of Population I, there is also a maximum number of emission nebulosities; in the outer arms, where dust, gas, and Population I fade out, the number of H II regions diminishes rapidly. The data agree very well.

Since the ingredients of an emission nebula are hot stars of sufficiently high luminosity (O and B stars) and the presence of dust and gas, the concentration of the H II regions in the spiral arms is easily understood. The correct interpretation is certainly that the spiral arms contain O and B stars and dust and gas. It is also clear why emission nebulae are not scattered everywhere: the exciting stars are absent outside the spiral arms. But is there dust and gas outside the spiral arms? There we cannot rely on the emission nebulosities, and need another argument.

The globular clusters of M 31 furnish the evidence on the presence of dust and gas outside the spiral arms. We shall see later that these globular clusters have a spatial distribution quite different from that of the emission nebulosities. We have every reason to believe that the gas is strongly concentrated toward the central plane, as in our own Galaxy, whereas the globular clusters are distributed in a spheroidal system. There are several hundered of them (probably 300 to 400), and this means that half of them are behind the central plane of M 31. Are they reddened or not?

It so happened that my exposure times (1 hour on blue plates and 4 hours on Hα plates) were so adjusted that a globular cluster was of the same brightness on both plates. A globular cluster on the far side of the system, which was projected far from the points to which I could trace the spiral structure, would serve as a test. For such clusters the difference of brightness on blue and red plates was practically neglible; my plates were so adjusted as to give equal densities at color index around $+0^m.6$. This opened the way to a very nice test of how these globular clusters behaved in regard to absorption. I made sure that there were quite a number of globular clusters lying outside M 31, where there were clearly no absorption effects (except for absorption between us and M 31, which we know to be very small). The problem was now to examine globular clusters all over M 31 for absorption. I could have detected a color difference of $0^m.2$. To my surprise it turned out that in general there was no effect whatsoever; wherever I found a globular cluster its color was the same. The faint globular clusters had magnitudes of 17^m or 18^m, and could always be distinguished easily from the occasional E galaxies, which have intrinsic color indices of $+0^m.9$, plus the effects of further reddening due to the expansion of the universe.

In general there was no effect, but in 10 or 12 cases the globular clusters were exceedingly reddened. Every one of these globular clusters was located in a spiral arm, and was obviously seen through the spiral arm. In general, the light of the clusters is not reddened, except when a cluster lies behind the plane, and then suddenly reddening is observed. This result was thoroughly confirmed recently when Arp made a new survey of the novae in M 31. Just one nova showed severe reddening, and again it lay right in a spiral arm.

So we observe that the emission nebulosities inside the spiral arms are greatly reddened, but most of the globular clusters shine through free spaces. Therefore outside the spiral arms there must be very little dust, and consequently very little gas. This led me to believe that not only Population I but also dust and gas are concentrated in the spiral arms. When I first published this result at the Conference on the Dynamics of Cosmic Clouds in Paris, it aroused severe doubts, and nobody wanted to believe it. The argument was that Struve had determined the strength of the interstellar H and K lines statistically to distances as great as he could, and the strength always increased with distance. My answer was that this was because Struve was looking through our own spiral arm — the Perseus arm was not yet recognized — and at that time I coined the term "local swimming hole," which the spectroscopists refused to leave. Fortunately, 4 years later the 21-cm measures of the hydrogen distribution convinced even the doubting

Thomases that the spiral arms are real concentrations of dust and gas, and that between the arms the density falls to a very low value.

The idea of a uniform distribution of dust and gas started as an assumption. Struve's work on Harvard plates was perfectly right. The whole trouble lay in making an unjustifiable extrapolation. There was another argument: half the mass in our neighborhood is in the form of stars, the other half in the form of dust. In itself this result is perfectly true, even today. But when it is extrapolated over the whole Galaxy it leads to wrong results.

It became clear that dust and gas are concentrated in spiral arms. This was important because it put an end to all attempts at explaining spiral structure on the basis of the mechanics of individual stars. The question now was: since spiral arms are loci of Population I and also of dust and gas, which is the primary thing? Are the stars formed from dust and gas, or are the dust and gas furnished by the stars?

There was one argument that indicated immediately that stars are formed from dust and gas. After the work of Bethe and von Weizsäcker, it was pointed out by many astronomers that supergiants are of very high luminosity and accordingly have very short lifetimes, say tens of millions of years, and must continuously be replenished. Secondly, we knew very well at the time that new stars are being born in great quantities, for instance, in the Orion Nebula. That meant that the primary phenomenon in the spiral structure is the dust and gas, and that we could forget about the vain attempts at explaining spiral structure by particle dynamics. It must be understood in terms of gas dynamics and magnetic fields.

Practically no quantitative measures were made, but this

is an example of how many conclusions can be reached without them. Of course, now it will be necessary to tie all these things down to numerical data. Nevertheless, the survey of M 31 has already been very illuminating.

CHAPTER 6

DUST AND GAS IN GALAXIES

In chapter 5 I discussed the distribution of dust and gas in the Andromeda Nebula; we saw that the dust and gas are concentrated in the spiral arms, and constitute the primary constituent. The dust and gas make the spiral arms, and the supergiants of Population I that appear there and also outline the arms are a secondary phenomenon; they are formed in the spiral arms.

The discontinuous distribution of dust and gas in our own Galaxy met with severe doubts, because the picture based on the continuous increase of the strength of the interstellar lines with distance had led to the idea that there was a thin sheet of dust and gas, uniformly distributed throughout the whole Galaxy.

That the dust and gas in our Galaxy are actually concentrated in the spiral arms was first shown by Guido Münch. To detect a concentration of dust and gas in a spiral arm, we simply turn to a region along the galactic circle where we hit a maximum or a minimum of galactic rotation. Münch selected as a test the stars of the Perseus association around h and χ Persei. They lie in the Perseus arm, fairly close to a minimum of the galactic rotation; in the mean the radial velocities should be of the order of -50 km/sec. We know that our sun is on the inner edge of a spiral arm. We now observe the interstellar lines in the spectra of the B stars in h and χ Persei. The lines of sight from these stars should cross our spiral arm with its dust and gas, which should give an interstellar line of nearly zero velocity, representing the center of rest for the spiral arm in which our sun is. This line should be of rather constant intensity, because the cluster stars are so close to each other that their light will all pass through the same complex of clouds in our own arm.

Because the stars of the Perseus arm are all at a distance of about 2.5 kpc, their interstellar lines should be shifted by 50 km/sec relative to the first line; a second line should appear, 50 km/sec toward the violet. Of course this line would have a variety of intensities for different stars, depending on whether a member of the association, which is rather large, is on the near side or the far side of it. By observing from one and a half dozen to two dozen stars in h and χ Persei, Münch showed that this is exactly the case. All the stars show the zero line, corresponding to the dust and gas in our own arm, and also a line shifted by 50 km/sec (corresponding to the velocity of rest in the Perseus arm) with a variety of intensities. Münch presented this first conclusive proof at the Rome Congress of the International Astronomical Union in 1952. At the same Congress, Oort presented the first results on the spiral structure of the Galaxy by means of the 21-cm line, which confirmed that the dust and gas are concentrated in the spiral arms, with very low density between the arms. The work of Münch showed the distribution of dust and gas in a very striking and clear-cut manner.

So the dust and gas are the primary phenomenon of spiral structure, and the appearance of supergiants of Population I is a secondary phenomenon; they are formed from the dust and gas. This meant that the spiral structure was not to be explained on the basis of point mechanics, but on the basis of gas dynamics.

The distribution of the H II regions along the spiral arms of M 31 raised the question why they are not also found outside the arms. The answer is that stars of high luminosity are absent outside the arms, and so are the dust and gas. Evidence that M 31 is not an exception, but that this is the rule in spirals, is furnished by the emission nebulosities in M 81. The number of known emission nebulosities is smaller (about 300 to 400) because M 81 is more distant, but it evidently shows the same features, so this is clearly a very regular thing in such galaxies.

As I mentioned at the end of the last chapter, it was clear that the Population I stars in the spiral arms must be rather young. By contrast, star formation has obviously come to an end in the E galaxies, because the dust and gas have been used up. We could conclude that star formation had possibly come to an end quite some time ago; in any case, the Population II stars would be much older than the Population I stars. So the first thing that lay behind the distinction between the two populations was probably a difference of age.

Let us now turn to the general distribution in M 31, aside from the spiral structure. The resolution of M 31 was carried out essentially in the central region — not right in the center, but farther out, where the density was not too high. At first glance it seemed that these Population II stars were restricted to the center. But when I made the survey of M 31, it turned out that every time I took a very good plate, even very far out, I always found a substratum of Population II stars pervading the whole Andromeda Nebula. In general, the intensity distribution that is usually brought out by photometry shows a very high peak in the center, and a slow falling off on either side.

The region between M 31 and M 32 shows the presence of Population II. Another region, 1°.5 from the center of M 31, shows the density gradient of Population II stars petering out along the major axis, and illustrates how Population II pervades the whole of the Andromeda Nebula.

Later on, when the 200-inch came into operation, it was easy to check these things. I was able to determine the extent of M 31 along the minor axis; it turned out that one could follow the Population II with the greatest of ease to about 5′ from the center. It is very beautiful, because in this direction, on account of the tilt, there is a very steep gradient, and one can practically swear that at one point there are some stars, but 2′ or 3′ farther out no more can be seen. Now this distance corresponds exactly to the distance to which Stebbins and Whitford were able to trace the light of M 31 with the photocell; they observed the intensity distribution along the minor axis, and reached a surface brightness of about 27^m to 28^m per square second of arc at this distance. This was a very satisfactory situation: one can see individual stars up to the same distance to which the light of the system can be traced with the photocell.

It is much more difficult to trace the extent of M 31 along the major axis, because the ratio of the axes is about 1:4, and everything is four times farther apart. The minor axis has the advantage that the isophotes are crowded together, but on the major axis the gradient is exceedingly small and one has to go out a considerable distance. Many plates must be taken before this point is finally settled. The only thing that I can say at present is that the population is still going strong to a distance of about 2° on either side.

I should have settled this point better if a very curious fact had not appeared, of which I was not aware until I later took special plates with the 48-inch Schmidt. It has something to do with the well-known radio observation that in our own Galaxy the outer part of the hydrogen is turned downward from the plane in one direction, upward in the other. So in our own Galaxy the gas is very much restricted to a plane between ±8 kpc, but from there on it turns downward slightly on one side, upward on the other. I went out 2° along the major axis in M 31, where I found Population II still strong; the next step was at 2°25′, and there I had lost it. Finally it turned out that the Population II suddenly swirls off to one side. I took a 7-hour exposure at the 48-inch Schmidt with special filters, and on this plate I could see that the Population II turns off on both sides of the major axis: it leaves the axis in one direction at one end, in the other direction at the other end. What is behind this thing? We shall have to check the orientation in space, and find in what direction it points. It is quite possible that this is evidence of tidal action between our Galaxy and the Andromeda Nebula; I say this with all reserve, for I have not checked the geometric situation.

That in all these cases we are dealing with Population II was finally demonstrated beautifully by exposures taken with the 200-inch. On each of these plates one would usually catch one, two, or three globular clusters not too far from the center; in every case in which such a globular cluster appeared, it was beautifully resolved at the same time as Population II. So we were quite certain that we were dealing with the same population as that of the globular clusters.

When one looks at the Population II stars, 1°.5 from the center, on the original plates, one immediately gets the impression that the main body of M 31 is really these Population II stars of the disk. Mayall, when he had seen the plate, expressed it very neatly: the real cake is the Population II stars; the spiral structure is just the frosting on the whole affair. This has also been brought out by quantitative measures. Holmberg made tracings in all directions through M 31, and was able to take out the intensity due to spiral structure. He could measure the spiral structure by itself, and he could measure the whole light contributed by the disk. He found that more than 85 percent of the whole light of M 31 comes from the disk population, and less than 15 percent from the spiral structure. To put it in a different way, if the whole spiral structure of M 31 were removed, the total absolute magnitude would drop by something like $0^m.4$, because most of the light comes from Population II.

Another argument shows that this picture is essentially correct. The integrated color of M 31 is $+0^m.90$, which is quite in agreement with what I have just said. The mean value for E galaxies is of the same order. This means again that the contribution of the bluer stars in M 31 to the total light is not very important.

Let me turn to a question that needs further investigation: what is the actual ratio of the axes in the projected disk of M 31? Although it looks like a very simple problem, it has not yet been solved. Probably the best information that we have at the present time comes from the contour tracings made by Hiltner and Williams on a film of the Andromeda Nebula taken with the 18-inch Schmidt at Palomar. They obtained quite an elongated affair with an axial ratio of about 4:1. On the minor axis their value agrees very nearly with the value of Stebbins, and with the actual resolution of the Population II stars.

On the other hand, a number of photometric investigations in which the observers studied only the intensity distribution along the major axis and the minor axis, but otherwise worked very carefully, gave values in the region of 1:2.4, or even 1:2. The value 1:2.4 was obtained both by Holmberg and by Redman. This result is disturbing; clearly it is very difficult to determine the extent along the major axis because of the low gradient; one works close to the plate limit. It is very important to determine this ratio.

Let us now consider what is the real inclination of the main plane of M 31 to the line of sight. Of course the best determination depends on the H II regions. One simply takes, for each spiral arm, the distance from the center on the major axis and on the minor axis, because there is every reason to believe that M 31 is just like our own Galaxy and that the H II regions all lie nearly in one plane. In this case one obtains an axial ratio of 1:4.9, and this leads to an inclination of 11°.7 between the main plane and the line of sight. On this point, then, we are certain, but we are not certain about the axial ratio of the general distribution in the disk. I suspect that the trouble comes from the presence of some sort of halo superimposed over the disk. Along the minor axis one would then have a rather elongated form (perhaps 1:4) for the disk of Population II, superimposed on a flat halo connected with the central system.

I think that is the solution, but nobody knows: it is very important indeed to settle the question. I have tried to put enthusiasm into several photoelectric observers, but after making the first attempts they have usually given up. It is not easy; to get the limits with photometric accuracy you must be sure that the area in which you measure contains no star brighter than, say, 21^m. The galactic latitude of M 31 is $21°$, and it is surrounded by a very uniform star field; you have to take very long exposures to be sure that your area is free of stars. The latter procedure is slow; you have to use a large telescope and be sure that the zero point does not change. But I think it must be done photoelectrically, for I cannot believe that it can be done with sufficient accuracy with photographic photometry.

I believe that there is a halo in addition to the disk because the most distant globular clusters seem to extend beyond any reasonable distance for the disk itself. We know that the disk goes to $45'$; if we use $1:3$ for the ratio and this well-defined observational limit, we find that NGC 205 is completely within M 31. The axial ratio $1:3$ will be an upper limit; the disk will certainly be flatter. Along the major axis objects will be found indicating that you have to go farther out. The known globular clusters extend considerably beyond the disk, which means that there must be some kind of halo. But this halo is considerably flattened compared to the halo of our Galaxy; it is not spherical, otherwise we should find globular clusters within the spherical area. A flattened halo plus a disk could produce exactly the confusion in which we find ourselves at present.

This concludes the preliminary survey of M 31; the result was a true insight into the spiral structure. Our knowledge that the dust and gas define the spiral structure has thrown new light on a number of problems. For instance, if an *Sb* galaxy like M 31 contained no dust and gas, it would mean that spiral structure would not be able to form, and that there should be no bright supergiants. The curious thing is that such systems actually exist. The *Sob* system NGC 5838 has no dust and gas and no spiral structure, yet it is an *Sb*. And not only do such systems exist, but they are very prevalent in the dense concentrated clusters of galaxies. Therefore these systems must have something to do with the *So*'s.

Hubble had introduced the *So* type, in a way that was not easy to understand, in his talk at the dedication of the McDonald Observatory. I had great difficulty in understanding the matter, because he was trying to fit in *So* as a transition type between *E* galaxies and spirals, although he did not find it easy to show why there should be these transition types. When it turned out that one would expect systems with no dust and gas to have no spiral structure, we were back at the *So*'s again. I remember having many discussions with Hubble on the point; I was not clear what he meant by *So*. He always showed me a collection of curious systems which had a little dark ring in the center around the interior, and dark in the center; there was absorbing matter around the rings, but otherwise they were amorphous *E* galaxies or related to *E* galaxies. I now understand his argument: he wanted to connect the *E* galaxies with spiral structure, and thought that these rings might be the first signs of incipient spiral structure.

It turned out that Hubble always classified systems like NGC 5838, without dust and gas, as *So*. So finally I realized that by *So* he simply meant "no spiral structure," and finally he admitted that the class should at least contain systems where one would expect spiral structure, but did not find it. He still insisted on the great importance of his limiting group of galaxies that showed absorbing lanes inside them. He believed that he could connect the *E* galaxies with spirals by regarding these systems as showing incipient spiral structure, which we know today simply does not exist.

In the end I think it is quite clear that, if we introduce the class *So*, we should define it as the class of galaxies in which from their general form we should expect to find spiral structure, but which, contrary to our expectation, do not show it. I think that Hubble finally accepted this view.

In 1952 I presented the same facts as I have just given, concerning the prevalence in clusters of galaxies of these systems without dust and gas and without spiral structure, in a lecture course at Princeton. Spitzer immediately said that it was quite evident that such systems should occur, because in these clusters there will be large numbers of collisions. The mean free path for the stars is so huge that the systems can move freely without doing any damage; if the collision velocity is high enough the transfer of energy from one system to the other is very small, as far as stars are concerned. But for the dust and gas the situation is of course quite different; there a real collision will occur, with the result that when the two systems separate again the dust and gas will be left lying between them. What will happen to it is another question.

If one assumes a collision velocity of 5000 km/sec or more, as one would expect at the center of clusters like the Coma cluster, the gas is heated to a temperature of over a million degrees. Spitzer was exceedingly interested to find out whether it was possible for the mass of hot gas left between the galaxies to cool down rapidly enough to get rid of this energy and condense again. This question is still unsettled; it is simply touch and go. But in principle this was an obvious and valid explanation of the stripped galaxies in the condensed groups like the Coma and Corona Borealis clusters.

At the time, Spitzer used the old distance scale and computed that there must have been several dozen collisions for an average galaxy in a condensed cluster. The calculation must be modified to take account of the new distance scale, and the number of collisions will not be so high, but that is all to the good; the process was so efficient that by the original calculation every galaxy should actually have been stripped, whereas we know very well that there are quite a number that are unstripped, or only partially stripped.

But a new and exceedingly interesting problem has come up. How are we to explain the So galaxies that we observe in the general field? I have become really interested in this problem, because on checking up the number of such galaxies in the general field I should say that it is embarrassingly large. If we want to explain these systems by the same collision process, we encounter the difficulty that we know the chances of collisions in the general field to be vanishingly small. If we want to explain these galaxies by collisions we must go back to an earlier epoch of the universe when the density was higher. Here the problem gets exceedingly touchy. There were certainly earlier epochs when the chance of collision was much higher, but, if the galaxies consisted only of gas at that time, they would simply have smashed each other up. If the field galaxies are to be explained on the basis of collisions, these were restricted to a very critical interval, after the majority of the stars had been formed. We know that in clusters of galaxies the luminosities reach up to the highest values, which is understandable, for if we take out the spiral structure, whatever we do, the resulting difference in luminosity is small. So they are practically as bright as other galaxies. And it seems absolutely impossible to squeeze all collisions between field galaxies into the same time interval. I have no definite results, but I know one thing: the number of stripped field galaxies is too large to be easily explained by the collision process.

There is another fact that cannot be explained by collision. In any case of a stripped galaxy, it should be very easy to find the other partner in the collision in the neighborhood today. If you take any reasonable collision velocity and let the systems separate with that velocity indefinitely, the two systems should still be on the same plate, even if they are nearby. I have searched the Schmidt plates for two or three dozen systems that I know are exactly of this type, and I cannot find the partner.

This problem has now become really interesting, and I shall return to it in another connection later. I see no way out other than to assume simply that these systems have used up their dust and gas in star formation, not in collision. I hope someone will take up this very interesting problem; it will be necessary to get a simon-pure list of systems that have been checked by short exposures with the larger telescopes, to show that there is no dust in them. Without such a list to start from one would certainly be lost.

CHAPTER 7

OUTLINE OF STELLAR EVOLUTION

In the last chapter I pointed out that the presence of dust, gas, and supergiants in Population I, and their complete absence from Population II, gave the first hint that the difference between the populations was very likely an age difference. All that could be said was that Population II was probably old compared to Population I.

The final clue would clearly be provided by the color-magnitude diagrams of globular clusters. Shapley's work had covered only the upper parts of these diagrams, because precise magnitude sequences were available only down to $16^m.5$ at the time; moreover, the photovisual plates were so slow that exposures were five or six times as long as for photographic plates. Good photovisual plates became available in the early 1930's, but the problem of magnitude standards for faint stars remained.

The 1P21 photomultiplier cell, developed as a result of the war, made it possible for the first time to establish standards photoelectrically down to the same faint limits as could be attained by photography with the same telescope. By 1950, when the 200-inch had come into operation, the time had come to attack the problem anew. Previous work on an object like a globular cluster had depended on making a large number of tedious and time-consuming transfers of magnitudes from distant standard areas. Now it was possible to set up the primary sequences in two colors within the cluster itself.

The problem was undertaken by Sandage, and his first studies (which were not on the U,B,V system, but I think were close to the International System) showed a well-defined giant branch, going over into the horizontal branch; every star in the gap of the horizontal branch is a cluster-type variable. Below the giant branch is the so-called subgiant branch, and below that the main sequence. Above the turnoff of the main sequence there is a faint sprinkling of stars, at about $+3^M.5$, which obviously lie very close to the main sequence. This feature has not so far been found in other globular clusters, and may be a peculiar characteristic of M 3. In general, the bulk of the stars set in at the turnoff from the main sequence.

The numbers of stars shown in different magnitude intervals in the diagram do not represent the true distribution. Actually, if all the stars on the main sequence had been plotted in the correct proportion to the giant stars, they would form a solid black area. The number of stars increases for fainter magnitudes, so the appearance of the diagram is deceptive.

I believe that George Gamow was the first to suggest the interpretation of the diagram that is accepted today. Shortly after the publication of my paper on the two stellar populations, I received a typical message from Gamow on a postcard: "Please tell me where the lower branch of the color-magnitude diagram joins the main sequence, and I will tell you the age of your Population II stars." I could only answer Gamow by telling him that nothing was known, that we intended to determine it as soon as possible, and that in the meantime he could extrapolate as he pleased. I promptly received a second postcard from Gamow: "With due respect to Schönberg and Chandrasekhar I have extrapolated the lower branch thus [there was a picture in which he had marked the point at which the subgiants turn upward]. O.K., four to five billion years."

This was a guess, of course, and that Gamow hit it so well was an accident, but his remark really contained the whole story of the interpretation of the diagram. The position of a star in the color-magnitude array is determined by mass and chemical composition. The composition changes on account of the energy production — the conversion of hydrogen to helium. Thus the position of the star is determined by its mass, its original chemical composition, and its age.

The members of a cluster of stars were all formed at the same time, and from the same huge lump of cosmic matter; in other words, they are stars of the same age and the same original chemical composition. We know that, after contracting from the protostar stage, a star finally reaches the main sequence. The mass-luminosity relation shows that stars of highest mass are also of highest luminosity. In any rich cluster we find that there are few stars of high luminosity; their number increases as one goes to fainter stars, probably reaches a maximum at a certain low luminosity, and then declines again.

Theoretical investigations have shown how the position of the main sequence is affected by the original composition. If the metal content is reduced to practically zero, the position of the main sequence lies about a magnitude below the average for stars in our own vicinity. If the abundance of light elements, including helium, is reduced to zero as well as that of the metals, the main sequence lies even closer to that of the nearby stars. If we take the stars of our own neighborhood, out to 5 or 10 pc, the data are excellent and are representative of K dwarfs and M dwarfs. These stars are certainly of all ages, and a variety of initial compositions, but the observed sequence is a very narrow one; as photometric measures are improved it seems to be growing still narrower, if we exclude a few subdwarfs.

So the observations fit very nicely with the theoretical computations, which means that the extreme cases of very low metal content, and so on, do not occur in nature. Otherwise the observed main sequence would not be as narrow as we observe it to be. We infer that the main sequences of different clusters will not differ very much on account of differences in the original composition.

The moment a star reaches the main sequence it begins to burn hydrogen, and in consequence its chemical composition begins to change slowly. Finally there comes a time when the star will have to leave the main sequence because the change in its composition has become so large. Theoretical considerations show that what happens to a star at this point depends on whether or not it is completely mixed inside. I think that Strömgren was the first to compute the evolutionary path of a thoroughly mixed star, which is not very interesting observationally: the star will slowly brighten, but will stay close to the main sequence until it has exhausted practically all its hydrogen, at which time it will turn up sharply to the left. But from the color-magnitude diagram of stars in our neighborhood and from the globular clusters we know that the left-hand side of the diagram is essentially free of stars, whereas on the other side there are plenty of stars — the supergiants and giants.

The computations for completely unmixed stars were first made by Schönberg and Chandrasekhar, in the paper to which Gamow referred on his postcard. For stars with masses greater than 1.2 solar masses (in which we are interested at present) the energy production involves the carbon cycle, and takes place in a convective central core which contains about 10 percent of the star's mass. Hydrogen burning is going on in this central core, and, since there is no mixing, the whole burning process is restricted to the core. Hydrogen is depleted and replaced by helium; finally an essentially helium core is left, and this helium core is isothermal because it contains at the time no active energy sources.

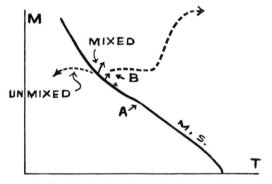

FIG. 2. Paths of mixed and unmixed stars (schematic).

The path of an unmixed star is shown in Fig. 2: it originally reached the main sequence at A; it moves essentially to the right, contrary to the behavior of a completely mixed star. At the point B, the whole hydrogen core, 10 percent of the star's mass, has been converted into helium. After this point the convective core begins to contract, and the computations of Schwarzschild and Hoyle show that the star then moves rapidly up into the giant branch. As a consequence of the contraction of the core, the temperature finally grows so high that helium burning begins.

For age determinations it is customary to take the point B as the starting point, because it marks the stage at which 10 percent of the mass has been burned in the form of hydrogen. If the mass of the star is known, say from the mass-luminosity relation, the computation of Schönberg and Chandrasekhar shows the initial position of the star on the main sequence. Thus it is possible to determine the age of a star that has reached point B:

$$t_e = \frac{M_*/M_\odot}{L_*/L_\odot}.$$

The age of a cluster like M 3 may thus be found to be 5 to 6 × 10⁹ yr, but there is an uncertainty at the present time of the order of 1.5 × 10⁹ yr, so it is only a first rough figure.

The formula gives ages for typical stars of the main sequence (Table 3).

Table 3.   Stellar ages.

Spectral type	$M_*/M_\odot$	$L_*/L_\odot$	t_e (yr)
O7.5	35	80,000	2.3×10^6
B0	16	10,000	1.2×10^7
B5	6	600	7×10^7
A0	3	60	6.5×10^8
F0	1.5	6	1.8×10^9
G0	1	1	7×10^9
K0	0.8	0.4	1.4×10^{10}

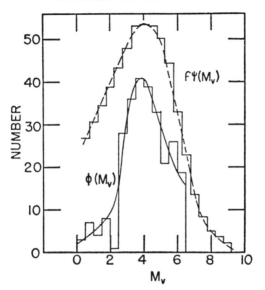

FIG. 3. The observed luminosity function for M 67 is shown by the solid line. The Salpeter original luminosity function corrected for the effect of the escape of stars of small mass is shown by the broken line.

Now we can see the role that is played by the luminosity function. If we could observe a star cluster from the beginning, and watch what happened to the main sequence, we should see it begin to disappear, first rapidly, then more slowly, until in the region of the globular clusters it is present only up to about $+3^M.5$. That was Gamow's argument: the whole main sequence has disappeared down to $+3^M.5$, and at that point the stars are now turning off.

Let us see why the globular clusters show the phenomenon in such a spectacular way. Originally they must have had a luminosity function like that shown in Fig. 3. But the bright stars have disappeared, and finally the top of the main sequence is represented by a point at which, for a small magnitude interval, the number of stars is very large. This is the reason for the difference between the color-magnitude arrays of globular clusters and open clusters. The open clusters still have stars on the upper part of the main sequence, which is poorly populated; only a few stars have turned off and we have little chance to observe them in the intermediate stage after they have left the sequence. For the globular clusters the stars are moving off in such huge numbers that the whole course of development can be traced; we can deduce the original luminosity function by extrapolating that observed for the fainter stars that are still on the main sequence. The evolutionary picture that I have sketched fits in very well with the observations.

The most extensive computations have been made for unmixed stars of about 1.5 solar masses, which represent the stars in globular clusters above the turnoff point. I have mentioned that Schwarzschild and Hoyle, guided by the color-magnitude diagrams, were able to trace the subsequent configurations of the star up to the point where helium burning sets in. From there on it is very difficult, and has not been done. In order to reproduce all the kinks in the giant and subgiant branch and to get it in the right position, Schwarzschild and Hoyle had to assume that the Population II stars much as those in M 3 have a very much higher hydrogen/metal ratio than the sun. Whereas the hydrogen/metal ratio is of the order of 8000 for Population I stars, the figure for Population II stars is 135,000, a ratio in metal content of about 17:1. This was one of the free parameters in the computation, and after it had been chosen it fixed the rest.

At the same time as they computed the evolutionary track of a globular-cluster star, Hoyle and Schwarzschild

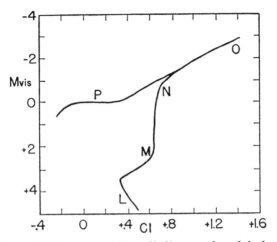

FIG. 4. Schematic Hertzsprung-Russell diagram for globular clusters.

computed the path of a star of the same mass and luminosity, but with a hydrogen/metal ratio 17 times smaller. They
found that the upper limit of luminosity was $0^M.0$, instead
of $-3^M.0$ as for the globular-cluster stars (see Fig. 4). The
difference is simply due to the higher metal abundance:
the greater electron abundance leads to higher opacity.

The case of high metal abundance is very important,
because it closely represents the open cluster M 67, which
is about the same age as a globular cluster, but of different
original composition. The upshot of the computations is
that the globular clusters are old, since their main sequences have been used up as far as $+3^M.5$, but in some
open clusters such as the τ Canis Majoris cluster the main
sequence begins at -8^M, in others at -5^M, and so on. For a
cluster like the Pleiades, where the sequence begins at
about -3^M, we cannot infer that the upper part of the
sequence has moved away, unless we can find stars that
correspond to it in the red-giant area of the diagram. Other-

wise it may be that no masses were formed in the cluster above those that correspond to the observed top of the main sequence.

Our conclusions refer to masses about 1.5 times that of the sun, but they should not be extrapolated to higher masses. The theory shows that the result applies to stars of less than 1.5 solar masses, if they are still using the carbon cycle. For higher masses we shall again have a convective core, which is finally transformed into an isothermal helium core; but Schönberg and Chandrasekhar found that this isothermal helium core cannot support the weight of a huge mass above it, and must contract rather rapidly. For lower masses, after an isothermal core has been formed, hydrogen burning still continues in the outer part of it, and the core becomes degenerate. But for stars with masses greater than 2 solar masses, the core is not degenerate; because of the rapid contraction in this case, the temperature in the core becomes so high that helium burning sets in at the center, in addition to the hydrogen burning at the surface of the core; the two processes go on at the same time. This has not been closely investigated, but it must be important for the giants and supergiants. Hoyle and Schwarzschild suspect that, as a result of the two processes going on there, a moment must come when the star has to throw off considerable mass in order to keep stable, and such mass ejection has been observed by Deutsch. A large amount of computation has still to be done, and even with electronic computers it will take a great deal of time. The only thing that we can say at present is that we feel that basically we are on the right track because of the work of Schwarzschild and Hoyle, but our age determinations are still very rough.

In the first studies of M 3, some part of the fainter magni-

tudes depended on extrapolation, but in the later diagram
this has been remedied, and the magnitudes are on the B,V
system (Fig. 5). The small number of stars in the lower

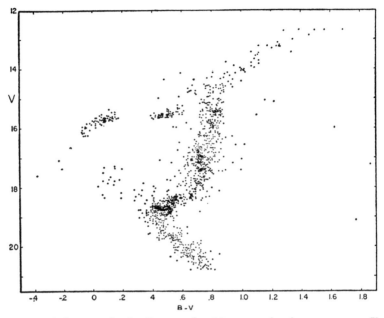

FIG. 5. Color-magnitude diagram for M 3 stars in the arguments V
and B-V.

main sequence, between $B-V = +0^m.4$ and $+0^m.8$, is of
course simply a consequence of the much smaller area
studied for these faint stars. Sandage went as far into the
cluster as he could, so as to include as few field stars as
possible. Field stars that are not cluster members would
set in, I would guess, at about $+0^m.3$, and would be scattered over the region of the diagram for which $B - V$ is
greater than $+0^m.3$. Statistically, these will be field stars,
though any individual could be a cluster member; the final

test is of course the radial velocity. The day will come when we shall be willing to spend the time necessary to measure the radial velocities of these stars.

The stars between V magnitude 17.5 and 18.5, with $B - V$ between $0^m.0$ and $+0^m.4$, could be well-mixed stars, which should move essentially along the main sequence. For a while this was a rather intriguing belief, because they are of about absolute magnitude $+2.5$ and spectral type F; and in the region just above the F stars we find the stars of rapid rotation, whereas further down the main sequence we no longer find high rotations. At the time it was thought that rotation provided the stirring mechanism, but the work of Mestel has since shown that rotation is not necessarily the stirrer. Strömgren's computations show that mixed stars will move upward several magnitudes, always staying close to the main sequence, and then turn off rapidly when the hydrogen content is practically exhausted.

All the red variables in M 3 lie on the giant branch, between $B - V = +0^m.8$ and $+1^m.8$ and between V magnitudes 12 and 14.5, and on the average they are completely intermixed with the stars of the giant branch. The long-period Cepheids lie partly on top of the gap, in a funnel-shaped area that is slightly inclined toward higher color indices at higher luminosities; they can be a magnitude brighter than the stars of the giant branch. We do not know where they come from; the only thing that is certain is that they are not associated with the horizontal branch.

We can be very certain that the exceedingly blue stars are members of the cluster, because M 3 is at galactic latitude 70°, and there is a rather negligible chance of finding a blue-halo star in this exceedingly sparse field. This, however, is a statistical argument, and the final proof will lie in the radial velocities.

It is quite possible that there may be some stars with color indices greater than $+1^m.7$ in globular clusters, but rich clusters like M 3 are really not the best selection for the search. If you want to do more, you need the luminosity function, which is exceedingly difficult to derive because of crowding; you can only study the outer regions, and extrapolation from these is very dangerous. It would be very much better to use one of the loose globular clusters, which have hardly any concentration to the center and are still sufficiently rich in stars; NGC 5053 would be ideal. This cluster has everything, including about 24 cluster-type variables; in the end I think we shall have to study it, to get a clear-cut case. There may be some very red stars in the center of M 3, but they will not be picked up because of the crowding of the central area, which contains probably half the stars.

The latest color-magnitude diagram for M 3 is reduced to mean points in Fig. 6; the circles represent photoelectric standards, not cluster members. If now we assume the present adopted value of $0^M.0$ for the cluster-type variables, and put in the main sequences for the Hyades and the Pleiades, we see that the two diagrams do not coincide. Even if the chemical compositions were the same, the curves could not coincide, because the globular-cluster stars are moving off the main sequence to the Schönberg-Chandrasekhar limit, and there is a real difference of slope between this part of the curve and the unevolved main sequence. This difference is, of course, independent of chemical composition; a change of composition would shift the curves vertically. The interesting test will come when we can finally push the diagram of M 3 down to the so-called zero position of the main sequence, and see whether it finally becomes parallel to the Pleiades and Hyades main

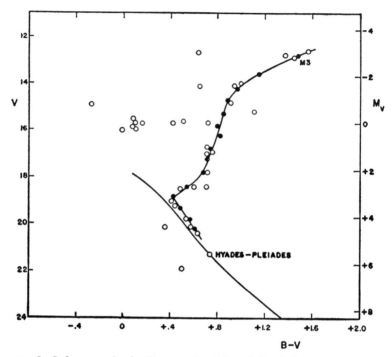

FIG. 6. Color-magnitude diagram for M 3 defined from the photo-electric standards (*open circles*) and from the mean color-magnitude diagram (*closed circles*). The main sequence from the Hyades and Pleiades stars is shown.

sequences. If the chemical composition of the M 3 stars is known with sufficient accuracy, this will give us the ideal determination of the zero point of the cluster-type variables. We are still very far from that, but it will come some day. Already one can see from the shift of the main sequence that the error in the zero point cannot be as much as a magnitude; probably it is not half a magnitude.

Now the stars on the main sequence, both of the Hyades-Pleiades and of M 3, conform to the mass-luminosity relation. But the stars in M 3 that have turned off are all stars of low mass, 1.2 to 1.5 solar masses, for we can ascribe them to their former positions. Thus we cannot apply the mass-luminosity relation to these evolved stars; to do that we need to know where they were when they lay on the main sequence. There has been a lot of confusion on this point among astronomers: it is not true to say that a star on the main sequence and a star on the horizontal branch at the place where it crosses the main sequence — and the two diagrams cross — are in identical states; they are not. They have the same color index and luminosity, but they differ in mass. A star on the horizontal branch of M 3 has a mass of about 1.4 or 1.5 solar masses, but a star on the main sequence, where this horizontal branch crosses it, has a mass of about 2 solar masses.

We should consider seriously the possibility of going from a two-dimensional to a three-dimensional color-magnitude diagram, by introducing the logarithm of the mass as a third coordinate. Our three coordinates would be the logarithm of intensity, logarithm of temperature, and logarithm of mass. The two-dimensional color-magnitude diagram is a projection of the three-dimensional diagram. I have made such a three-dimensional model for myself, and I hope that other people will be interested in doing it again, for it is very instructive. When projected on the three planes, it gives the mass-luminosity relation, the H-R diagram, and the mass-color relation. There is a very good chance that in the near future we shall recognize variety in the chemical compositions of stars; this would call for a fourth dimension in the model, which might lead to using different colors. At the present time we have need for only three dimensions.

CHAPTER 8

THE DISTANCES

TO THE GALAXIES

O U R survey of the properties of galaxies has led us directly to the problem of stellar evolution. Discussions of stellar evolution have been going on for a long time, but finally we have reached a point at which we can handle the problem. We can regard the form of the color-magnitude diagram of a group of stars of common origin as the result of changing chemical composition; the main features stem from the conversion of hydrogen into helium. The next logical step would be to examine our own Galaxy in detail and try to understand it in the light of this picture. But first I should like to devote a chapter to the determination of the distances of galaxies.

If we want to compare two galaxies, or two objects in the same galaxy, it is essential to determine their distances with the same yardstick. The Cepheids have provided this yardstick in the past, and will probably do so for some time to come, because they have the agreeable property that

one need only determine their periods and light curves —
spectrographic study is not necessary — and the whole problem reduces to the determination of their absolute magnitudes.

Historically, the first step was the discovery, by Miss
Leavitt at Harvard, that the mean apparent magnitudes
of Cepheids in the Magellanic Clouds were simply a function of period. This relation was promptly expressed in
terms of absolute magnitudes by Hertzsprung, who used
the parallactic motions of 13 Cepheids in 1913. In 1916
Shapley made the first application of this period-luminosity
relation, not to galaxies but to globular clusters, since he
was interested in extending the relation to the cluster-type
variables, the most frequent type of variable star in globular
clusters. Bailey, in his study of globular clusters, had found
about a dozen variables that looked like Cepheids of long
period; each cluster contained only one or two such variables. Shapley constructed a diagram by connecting these
variables with the cluster-type variables in the same clusters, and thus put together the upper part of the diagram.
By bringing this upper part into coincidence with Miss
Leavitt's diagram, he obtained a relation that included the
Cepheids and the cluster-type variables, and this period-
luminosity relation was the one accepted for the next 40
years. The validity of the step was not questioned at the
time, because very little was known about the Cepheids
in globular clusters; they were known to have light curves
in general like those of galactic Cepheids, except that for
periods between 12 and 20 days there was a somewhat
different form of curve — the so-called W Virginis type.
This, however, was hardly known at the time. Shapley redetermined the absolute magnitudes, omitting from Hertzsprung's list l Carinae, and also κ Pavonis (on account of

the variability of its light curve and period, practically unknown phenomena for Cepheids), now known to belong to Population II. As Shapley used practically the same stars as Hertzsprung had done, and also the same proper motions, he arrived (within o^m.2) at the same absolute magnitudes.

The determinations of the zero point by Hertzsprung and Shapley did not take account of absorption of light within our Galaxy. In about 1928, Trumpler's work showed that absorption was an important factor, and necessitated a new determination of the zero point, which was made at Mount Wilson in 1939 by R. E. Wilson. His material was not of the same high quality as that of the old Boss catalogue, but he used a much larger number of Cepheids, 157 in all. Most of the new material was from the new Boss catalogue, but he included a number of proper motions that he had derived himself, and in addition he had radial velocities for a number of Cepheids. He used the parallactic and peculiar motions, derived the solar motion directly from the radial velocities, and used the galactic rotation with Joy's value of 20.9 km/sec kpc for the rotation constant A. For the absorption he used Joy's statistical value of o^m.85 kpc^{-1}. The resulting correction to the old zero point was $-$o^m.14 $\pm$ o^m.2, so the change was not very large.

At the same time a number of determinations of the zero point for the cluster-type variables were made; these had the advantage that they were not much affected by absorption. Data were derived at Harvard by Mrs. Bok and Miss Boyd for seven cluster-type variables, and the material was extended to cover about 67 stars by data from Luyten and from French observers. The proper-motion data were critically discussed by Fletcher, and it turned out that the proper motions for very faint cluster-type variables were simply errors of measurement. The mean corrections to the zero point for the cluster-type variables turned out to be $+0^m.08$ from the v component of the parallactic motion, and $+0^m.07$ from the τ component, again indicating that the zero point of the period-luminosity relation was nearly correct. The result for the cluster-type variables inspired great confidence in the zero point, on account of the negligible effects of absorption.

Nevertheless there were some disquieting facts. I have already mentioned the discrepancy between the luminosities of globular clusters in our own Galaxy and in M 31: the upper limit was around $-9^M.0$ in our Galaxy, and barely $-7^M.5$ in M 31. This difference was very well established: Christie had determined the integrated magnitude for most of the globular clusters in Shapley's list, and in M 31 the upper limit has been checked by Stebbins and Whitford with the photocell. There was a discrepancy of $1^M.5$.

A similar difference appeared in the mean brightness of the novae in the two systems. In our Galaxy the excellent determination of Gratton and Cecchini gave a value of $-7^M.2$, but in M 31 the mean brightness was only $-5^M.7$. The novae should not be considered in the same class as the globular clusters, for Hubble's observations were exceedingly scattered, and the absolute magnitudes depended on the use of a standard nova curve to extrapolate the magnitudes to maximum. Nevertheless, the novae indicated that there was a discrepancy of the same kind.

Another serious problem was that, as knowledge stood at the time, our own Galaxy was much larger than any other, although here again the data were uncertain, because we could not then make a strict comparison between our

Galaxy and M 31 by using the same objects. The stars that revealed the greatest extent of our Galaxy were the cluster-type variables, which were of course inaccessible in M 31.

I have already mentioned the many attempts that were made to explain the discrepancy. I made a new determination of the magnitude scale, with the result that Hubble's distance modulus, $m - M = 22^m.2$, was revised to $22^m.4$; although there were still uncertainties in the magnitude scale, up to $0^m.5$, they could not explain the 1.5-magnitude discrepancy. Hubble favored the explanation that the brightest globular clusters in M 31 were $1^m.5$ fainter than the brightest in our Galaxy, though it was difficult to understand why that should be, since the samples were about equally rich in both systems. Another possibility was that these objects in M 31 were not globular clusters. They had not been resolved at the time; the only thing that we knew was that their spectral types were F5, F8, G0, like those of globular clusters. There the matter stood in 1939; none of the explanations was convincing.

Then, during the war years, the concept of the two populations emerged. There were two kinds of Cepheids, one belonging to each population; were they really the same kind of star? There was no reason any more to suppose so, and there was considerable doubt whether the Cepheids of the two populations had the same luminosity. This permitted us to consider whether the unified period-luminosity relation should be separated again. Probably a decision could be reached by means of M 31, with both populations side by side: the brightest stars of Population II had already been revealed with the 100-inch. We looked forward to using the 200-inch to decide whether the two period-luminosity curves should be separated.

I remember the first plates I took in 1950 with the 200-

inch; on blue plates with a correcting lens and a 30-minute exposure I could just reach the brightest stars of Population II. Under fine seeing conditions, with a 30-minute exposure, one could see the whole area, especially outside the nuclear area, peppered with these faint stars, all very close to the plate limit. That meant that there was no hope of reaching the cluster-type variables. We had expected to find them at $22^m.4$ — the 30-minute exposure with the 200-inch would certainly reach that magnitude — and we got only the brightest stars of Population II. I mention this especially because it is occasionally mentioned in the literature that I actually searched for cluster-type variables. I never made such a search, because it was quite clear that I was reaching only the brightest stars of Population II, which are $1^m.5$ brighter photographically than the cluster-type stars.

The problem now was to make a really good determination of the limiting magnitude that we could reach in 30 minutes with the 200-inch. That took some time, because new sequences had to be established with photocells, including S.A. 68, the comparison area for M 31. By the fall of 1952, Baum gave me the first sequence in S.A. 68, extending to 22^m, and I was able to extrapolate from it to the limit reached in 30 minutes — about $22^m.75$ to $22^m.80$ on the plates I used at that time. Later observations thoroughly confirmed this value: the brightest Population II stars in M 31 had photographic magnitude $22^m.75$.

In the meantime it was also possible to derive the distance modulus of M 31 from the classical Cepheids, using Hubble's data and the new magnitude scale: the value was about $22^m.7$. And the brightest Population II stars were $22^m.75$. At the same time I had resolved all the globular clusters in M 31 into stars, and the magnitude at which this happened was $22^m.75$. On red plates I resolved them easily, because I could give exposure times up to 2 hours; on blue plates, under the best seeing conditions, I could just detect incipient resolution with 30-minute plates. So it was clear that the stars in the globular clusters and in the general field were the same. And, since we knew that their absolute magnitude was $-1^M.5$, the distance modulus of M 31 must be about $24^m.25$, whereas the distance modulus derived from the classical Cepheids, on the correct magnitude scale, turned out to be $22^m.7$. Actually, my old scale, determined with the platinum half-filter, had an error or $0^m.3$ at $22^m.4$; when this error was taken out, the distance modulus from the Cepheids became $22^m.7$.

We now had two determinations of the distance modulus of M 31: that based on Population II was $24^m.25$; that based on the Cepheids of Population I was $22^m.7$ — again a difference of about $1^m.5$. This meant that in the period-luminosity relation the cluster-type variables must be shifted downward by $1^m.5$ relative to the classical Cepheids, or the classical Cepheids upward relative to the cluster-type stars. There was actually a physical difference, which showed up when we could observe the two populations side by side.

Which of the two distance moduli should we accept? Either one is right, or the other is right, or both are wrong. The decision I made at the time was that the safer of the two values would be the one that depended on the cluster-type variables. First, the proper motions of cluster-type variables are quite large — quite a number of them are in the first decimal of a second of arc — whereas those for classical Cepheids are all in the third decimal. Secondly, absorption plays a decisive role for the galactic Type I Cepheids (as we shall see in more detail later on) but is hardly significant for cluster-type variables. Third, the absolute luminosity of the cluster-type variables seems to be especially well determined; we could probably eliminate those of thirteenth magnitude whose proper motions were simply errors of measurement. And even if there should prove to be subgroups among the cluster-type variables, the fact that we had both proper motions and radial velocities for each star strengthens the result for the mean absolute magnitude. So I concluded that probably the correct value for the distance modulus was $24^m.25$, and that the error lay in the adopted zero point for the classical Cepheids.

When I announced these results at the Rome meeting of the International Astronomical Union in 1952, Thackeray rose and stated that he had just found the first cluster-type variables in one of the globular clusters in the Small Magellanic Cloud, and that they showed up not at $17^m.4$, as they should have done according to the old period-luminosity relation, but at $18^m.9$. They were fainter by $1^m.5$. This of course confirmed very nicely the result that had been obtained in M 31. But of course it did not indicate which zero point was the correct one. We simply knew now that the two curves must be separated by something of the order of $1^m.5$.

The exact value of the relative shift of zero point can be most easily determined in the Magellanic Clouds. The photometric scale for the Clouds must be checked; it probably needs correction even down to 17^m, and the extrapolation down to 19^m is quite uncertain. The work is now under way; Arp has already done part of it with the 74-inch in Pretoria, and Sandage is now working to determine the difference. Shapley and Mrs. Nail obtained about the same value, $1^m.5$ or $1^m.6$, using an extrapolated magnitude scale. There should be no difficulty in determining the difference to within $0^m.02$. The problem of which zero point is the correct one is, of course, still unsettled. If we use classical methods, we have two choices in solving the problem: we can determine the zero point of the classical Cepheids, or we can use the cluster-type variables. I think we must realize that in the foreseeable future we cannot hope to reduce our errors below $0^m.2$; if we want to do better, we must use better methods. Such methods exist, but they need to be improved. One such method is to determine pulsation parallaxes for Cepheids; another, which has opened up recently, depends on the discovery of a few Cepheids in galactic clusters.

An attack on the problem by classical methods was made by Mineur in 1944; he derived zero points separately for the cluster-type variables and the classical Cepheids, but he was so much under the influence of the old period-luminosity relation that he simply gave the mean as his final result, which led to a correction of $-0^m.8$. It would have been wonderful if the whole question had been settled by Mineur's paper, but this was not the case.

Mineur used essentially the same 167 classical Cepheids as Ralph Wilson, who had derived a very small zero-point correction. Mineur realized that he must be very careful about the plane of the Galaxy, because of the strong concentration of the Type I Cepheids to the galactic plane, and of the importance of absorption. So first he redetermined the plane of the Galaxy from the 167 Cepheids, and got a value that is in excellent agreement with the latest determinations, until the modern radio values.

As a means of handling the absorption, Mineur used the idea that the mean z component for the Cepheids should not change with distance, an idea proposed many years ago by Bottlinger. This means that all Cepheids lie very close to the plane of the Galaxy. So he divided them into groups, the criterion being that there should be no change in z with distance — a very sound idea, as we know today. But his final solution was loaded with many unknowns, such as the absorption, and the coefficients of some of these unknowns in the equations of condition were rather small. So although his value for the classical Cepheids agrees very well with the modern values, the coincidence must be considered spurious.

About two years ago, Blaauw and H. R. Morgan made a new determination of the zero point of the classical Cepheids which gives every evidence of being correct. Morgan had derived proper motions of very high accuracy for a limited number of Cepheids, partly the same as the lists of Hertzsprung and Shapley. Curiously enough, although the accuracy is very much higher — the fourth decimal is added to the proper motions, although it is probably largely fictitious — these proper motions do not differ very much from those of the old Boss catalogue.

Now comes the allowance for absorption. I shall take only the solution made by Blaauw and Morgan for the stars for which Eggen had made direct determinations of color excess, eight stars in all. These stars led to a zero-point correction $\Delta M = - 1^m.4 \pm 0^m.4$. They made other solutions, but they had to make an assumption as to the amount of absorption per kiloparsec, and we know that such assumptions are not very realistic, because of the discontinuous distribution of matter in the spiral arms. Eggen's color excesses need a correction, because Eggen simply assumed that the nearest Cepheids, δ Cephei and η Aquilae, are not reddened at all, and we know now that this assumption was wrong. A reddening of $0^m.1$ raises the absorption by $0^m.4$, and the necessary correction will increase the value of ΔM. I mentioned that the difference between the new proper motions and those in the old Boss catalogue for the same stars is not very large. This means that the whole trouble in this determination arises from the absorption. We have not licked it yet, and we shall not lick it until we have the true color excesses for these stars, based on the true intrinsic colors for Cepheids of different periods. When the color excesses have been determined, Blaauw intends to repeat the solution.

Blaauw's solution has several points in its favor. For one thing, if you resolve the proper motions into a v component (in the direction of the apex) and a τ component, the motions of all these Cepheids, almost without exception, point in the direction of the antapex, and this although all the proper motions are so small that they start only at the third decimal (for instance, $0''.004$); this shows how good these proper motions are. This was not the case for the old material used by Ralph Wilson and by Mineur.

Another point which inspires confidence is the comparison of the average transverse motion of the Cepheids with the average radial motion: $|T| \approx 8$ km/sec, and $|R| \approx 9$ km/sec — exactly what one would expect for stars which have very small motions in the galactic plane; the values agree. We can compare these values with those for G supergiants of the same luminosity in our neighborhood. Hill has determined $|T| \approx 9$ km/sec and $|R| \approx 10$ km/sec for G supergiants of the same luminosity as the Cepheids.

These criteria show that the propor-motion data used by Blaauw and Morgan are thoroughly reliable, and also that the absorption has been taken out to the point where it causes no difficulty. For the G supergiants the color excesses are actually known.

A new determination of the zero point of the cluster-type

variables is being made by van Herk, with the plates which van Maanen obtained in 1919–20 with the 60-inch Cassegrain reflector, which have been repeated in the meantime. In a year or two we shall have the results, a new determination of the zero point based on the cluster-type variables in our Galaxy. Of course you get accurate proper motions in 30 years at the Cassegrain focus. The radial velocities have been determined for all his stars; if there are any subgroups they will be taken care of. Moreover, Kron at the Lick Observatory has determined the color excesses for all these cluster-type variables; although the color excesses may be small, and of importance only for variables near the galactic plane, we shall have a determination that is corrected for them.

This is as far as we can go in determinations of the zero point by classical methods. Even with the best data we cannot expect to bring the uncertainty very much below $0^m.25$; we shall have to turn to other methods.

For the determination of pulsation parallaxes for Cepheids there are accurate photometric data; the only difficulties at the present time are with the temperature scale. At present it is very difficult to derive the effective temperature from the observed colors. But the prospects for the future are bright, because photoelectric scanners are coming into operation. At the Washburn Observatory, Code has a program of photoelectric spectrophotometry of Cepheids by scanners, from λ 10,400 to the far ultraviolet. The results that he has obtained for other stars make it very clear that if this method does not led to an accurate determination of temperature we shall never succeed in determining pulsation parallaxes. But this would be an entirely independent method of determining the zero point for the Cepheids.

The possibility exists of determining the absolute magnitude of a Cepheid in a galactic cluster by determining the color-magnitude diagram for the cluster stars. At first this looks very promising, but it turns out that most of these clusters contain few stars. Moreover, in many cases it is difficult to go faint enough to establish the zero main sequence; the stars that can be observed are already turning off and moving away from the main sequence. Nevertheless, these results should furnish a valuable check, and probably will be of the same order of uncertainty as we should get if we had perfected our classical methods.

The presence of Cepheids in clusters has been known for a long time; one of the Cepheids known today to be in a cluster was used in 1925 by Peter Doig to determine the distance of the cluster.

According to a paper that I have just received from Irwin, there are ten known Cepheids in clusters, and though not all have been checked for radial velocity they are very likely members. After the Palomar Sky Survey became available, a number of observers simply looked at each Cepheid on the corresponding map to see whether it was in a cluster, and quite a number were found in this way.

At the present time, work has been finished on four such clusters (Table 4).

In the cluster NGC 7790 there is another Cepheid, CE Cas, which is itself a double, consisting of two Cepheids which can easily be photographed separately with the 200-inch telescope. They are rather more than 1″ apart, and short exposures are needed, because they are so bright. I have taken a number of plates of the pair in recent years for Sandage, who is studying it. So NGC 7790 should finally furnish three Cepheids, but the light curve is at present available only for CF Cas.

Table 4. Period and absolute magnitude
of Cepheids in galactic clusters.

Cepheid	Cluster	Period (d)	Absolute magnitude (B)
EV Sct	NGC 6664	3.09	−1.9
CF Cas	NGC 7790	4.87	−2.5
U Sgr	M 25	6.74	−3.1
S Nor	NGC 6087	9.75	−2.8
			Mean −2.6

The absolute median brightness on the B system is given in the last column of Table 4. It follows from the color-magnitude diagram, to which the unevolved main sequence has been fitted. Although the method at first looks very exciting it is observationally very difficult, and the color-magnitude diagrams show how precarious the fit undoubtedly is. The individual accuracy is not very high. In all the clusters, the upper part of the main sequence is already turning off. The allowance for absorption is determined by means of the standard reddening diagram, and the absolute magnitude given in the table is derived. For NGC 7790, which contains CF Cas, the fit of the main sequence is better than in the other two cases, but it is still precarious; a shift of $0^m.2$ is still possible.

These data are on the B system; to go to the International System we add a correction of $-0^M.2$, which would make the mean $-2^M.8$. The value, taken from Shapley's curve, that corresponds to the mean period of these Cepheids is $-1^M.5$. Thus the correction to the zero point of the period-luminosity relation would be $-1^M.3 \pm 0^M.26$ m.e.). So we are in the same range again. But it is evident that there are considerable difficulties here.

Of course in a very rich cluster all these difficulties would disappear. Such a cluster is NGC 1866 in the Large Magellanic Cloud, a very unusual cluster, in which the brightest stars are of the order of -3^M, and so rich that everybody who looks at a photograph swears it is a globular cluster. This is a type of cluster that we do not have in our own Galaxy; its absolute magnitude is $-9^M.2$, which comes up to the limit of the brightest globular clusters in our own Galaxy. Shortly before the war I asked Thackeray to shoot NGC 1866 with the 74-inch in Pretoria, because it should be very simple to decide whether it is a globular or an open cluster.

Assume that it is an open cluster and has giant stars. You simply take a blue and a red plate. On the blue plate, select the brightest blue stars that you can find; and select also the brightest red stars. This is the trick: if it is an open cluster, the two groups will be equally bright, but if it is a globular cluster, the brightest red stars will be $1^m.5$ brighter than the brightest blue stars. Thackeray made the test and swore again that the cluster looked to him like a globular cluster, but this criterion showed that it could only be an open cluster: the brightest blue stars were if anything a little brighter than the brightest giants. This made the cluster especially interesting.

After Thackeray had found cluster-type variables in some of the Magellanic Cloud clusters, and after Gascoigne and Eggen had separated the clusters into two groups, one with color indices near 0^m, the other with color indices near $+0^m.5$, or $+0^m.6$, it became clear that the latter group most likely contained globular clusters — globular clusters in our Galaxy have similar color indices — and that the former were open clusters. The color index of NGC 1866 is $+0^m.07$, in the group with small color indices. A search for cluster-type variables (because it looks like a globular cluster) had resulted in the discovery of about ten classical Cepheids. There is no doubt that they are all members of the cluster, and their periods are all close to 3 days. Sandage is working on the color-magnitude diagram of the cluster. The problem of course will be whether, in this rich field in the Large Cloud, it will be possible, in the ring area around the central burned-out core of the cluster, to separate the cluster members sufficiently from the background, which is unfortunately very rich. If we cannot separate enough cluster members, we shall simply need a larger telescope. I fear very much that this is the case, as it was with two smaller clusters — the ring area was so small that one could not sort out the cluster stars. We hope that NGC 1866 will furnish a useful diagram, of sufficient accuracy to tie in the Cepheids with an error of the order of $0^m.1$, or even less.

I have been stressing the ways in which we could improve the determination of the zero points. But assume that by sheer good luck we suddenly reached accuracy, that we were sure of the zero point to within $0^m.05$. Then there would be a terrible embarrassment, because we could not fully utilize it. The reason is that at the present time we cannot establish our magnitude sequences with this accuracy for galaxies. In the Magellanic Clouds perhaps we could do it; perhaps in the near future we shall have a 100-inch telescope in the Southern Hemisphere. But in M 31 and M 33 we simply could not utilize such accuracy because we cannot establish our primary sequences in such galaxies with this accuracy. The reason is the faint background of unresolved stars; it presents real problems, which were brought home to me when I tried in recent years to establish accurate sequences in M 31.

The normal procedure that we have followed up to now is to select an area in the sky in which our sequences are determined photoelectrically, say down to 23^m, which can be done today with high accuracy. Now to transfer this field to M 31, you expose half your plate with a given exposure time to the area of M 31 in which you are interested; you then turn your plate around and expose the other half to the standard area. You are in trouble right away. The principal law of photometry is that two things are equal if they produce the same density on a photographic plate under identical conditions. Half the plate shows a field where you have only your stars, plus the illumination by the sky; the other half has the unresolved background in addition to the general sky illumination. When the background is exceedingly weak you may get away with it — the errors are small. But even so, when you can just see the background, the faint stars can be systematically off by as much as $0^m.25$ in spite of all precautions; I have made the actual tests. If the field intensity is higher, say close to the center of M 31, it is remarkable: errors of $0^m.6$ and $0^m.7$ can be made simply in transferring the scale.

You may say, why not establish the primary sequence directly in these fields photoelectrically? At first the photoelectric observers were quite eager to try it, but when they saw the scatter in their observations they flatly refused to go on, because they felt it below the dignity of their profession to be satisfied with errors of the order of $0^m.1$. Working in the field itself you have to search on a comparison plate for an area that can be used as an empty field — you have to subtract it anyway. The background fluctuations are already so large that you cannot do better than $0^m.1$. The situation becomes frightful if you work in the dense parts, toward the center of M 31; you have to be very careful because the photocell is so sensitive that you can feel the gradient in the field, and the moment there is some disturbance of the gradient due to obscuration, the whole thing goes wild. So it seems impossible at the present time, as far as we can see, to get any accuracy for an individual star, free from systematic error, which will be of the order of $0^m.1$. The fact that the background effect is so serious rules out quite a number of possibilities in determining the distances of galaxies, and we should realize it.

For instance, it has often been suggested that the novae would be among the most useful distance criteria. It would be a most painful thing to attempt; in M 31 the novae are anything but bright. One has only to read Arp's paper on the novae to see what trouble he had in getting magnitudes free from background effects. If it had worked in M 31, the next galaxy in which one would like to apply it would have been M 81, because in spite of every effort we have found only one Cepheid variable there. And one cannot believe that this is a galaxy with only one Cepheid. But in the center of M 81 there is a fireworks display of novae just as in M 31; actually it seems as if the nova activity in M 81 is even higher. But here is the problem: the brightest novae in M 81 all appear preferentially in the dense central lens. They are 3^m fainter than in M 31, which means that their maxima, instead of being at 15^m, 16^m, 17^m, are at 18^m, 19^m, 20^m. The surface brightness of the galaxy is the same as that of M 31; the fainter the stars, the larger the effect. This simply means that at present we can forget about novae as distance indicators. Novae have been observed in M 87, the largest members of the Virgo cluster, and I think it would be extremely dangerous to base a distance of the Virgo cluster on them, unless we put in a terrifically long time observing, and chance to find one so far out that the effects of background density are small. You could reach only the fast novae, and you would have to make daily observations, at the least, to be able to extrapolate properly to the maxima. We have to realize that from now on the greatest practical difficulty in determining distances will be the elimination of this background effect, and we are not too clear as to what will be the best way to do it.

So we should not worry too much at the present time if we cannot get more accurate determinations of the zero point. If we reach an accuracy of $\pm 0^m.1$ we have every reason to be satisfied; even if the accuracy were higher we could not use it at the present time. This is a very unpleasant fact, but it is better to face the music.

The situation is especially serious because it is clear that in the next few years we shall be at our limit with the existing instruments, and with all these unpleasant problems to contend with. What is the best distance determination of M 31 now? Probably for M 31 itself the ideal solution would be to go out very far from the center, where Population II is thinning out rapidly, and pick out a dozen of the brightest Population II stars; in photoelectric work you have only to watch that you do not hit some faint extragalactic nebula, which is easy to dodge because of its extent. Make the determination there; but even for that the measures would probably take two weeks.

The important thing is that we seem to be on the way to a consistent scheme; no contradictions between the distant galaxies have shown up at present. There are many important questions. Are the cluster-type variables everywhere of the same absolute magnitude? It is not settled at the present time. We should be able to check it in the Magellanic Clouds, because Thackeray and Wesselink have apparently found the first field cluster-type variables there. In order to settle such questions we need a consistent system of magnitudes; we can be satisfied today that it is within $0^m.2$, and it might be necessary to tie it down within $0^m.1$.

CHAPTER 9

GALACTIC CLUSTERS

W E H A V E seen that the color-magnitude diagram of M
3 provides a clue for understanding the diagrams of the two
populations, because in this case the main sequence starts
only at $+3^M.5$. We concluded that the brighter part of the
main sequence has disappeared long ago, and that at the
present time the stars are moving into the subgiant branch
at $+3^M.5$. This interpretation was supported by the theoretical computations of Hoyle and Schwarzschild, who
started from the computations already made by Schönberg
and Chandrasekhar and succeeded in representing the observed diagram very well. The age that they arrived at for
these phases was of the order of 5 to 6 $\times$ 10^9 yr.

For practical purposes it is convenient to identify the
phase at which the convective, isothermal hydrogen core of
the star is exhausted, and the track begins its horizontal
progress to the right. At this point the star has used up 10
percent of its mass in the form of hydrogen, and converted

it into helium. The progress of the star is shown in Fig. 2; at A it starts hydrogen burning; then it moves upward, and the point B, at which the track turns to the right, is defined as the terminal point of the main sequence. The point at which the convective core is exhausted can be used to compute the age of the cluster by means of the formula

$$\tau = 1.1 \times 10^{10} \ M/L_T \ \text{yr},$$

where L_T is defined as the luminosity at the termination point of the main sequence; M and L_T are in solar units. This formula gives the age from the time at which the star reached the main sequence. The time given takes account of the time taken in brightening from A to B, therefore the constant is a little larger than that given on p. 86, where the age was deduced from the initial point on the main sequence, which is a little difficult to ascertain, though you can do it if you make the computations à la Schönberg-Chandrasekhar.

These methods of determining the ages of the clusters are still very rough, simply because there are not yet enough theoretical computations. Henyey and his collaborators have computed the phases by which a star contracts from the protostar stage and reaches the main sequence: the star moves toward the main sequence essentially on a horizontal line, and shortly before reaching it, it drops down slightly. The moment it reaches the main sequence, hydrogen burning starts, and afterward the star moves off to the right as described before.

The formula refers to an original composition of 80 percent hydrogen, 20 percent helium. Actually the percentage of hydrogen by mass is what matters. Although a different primeval percentage of helium would change the result, the change will not be very essential so long as we stay within one population. And we shall see later that we have very good reason to believe that this young Population I is very homogeneous in chemical composition.

The galactic clusters furnish us with evidence on the evolution of Population I stars. The early studies of their color-magnitude diagrams were very strenuous affairs, because they had to be done photographically, and it was very difficult to keep the scales correct so that the color indices should not have systematic errors. Nevertheless, some remarkable results were achieved; Hertzsprung's early work on the Pleiades practically comes up to modern standards. Later came the work of Heckmann and Haffner on Praesepe, which is comparable to modern photoelectric observations, but required a terrific number of photographic plates.

Trumpler was the first who systematically made use of the Hertzsprung-Russell diagram in order to get information about star clusters. He worked on the problem all his life. The program was very ambitious: he determined magnitudes and color indices by photographic methods, and plotted the upper portions of the color-magnitude diagrams. He also obtained spectra for as many of the stars as possible, and measured the radial velocities. The work is now being prepared for publication by Weaver.

Altogether Trumpler observed 100 open clusters, and his method of classification reflects features that we consider today to have evolutionary significance. Class I referred to clusters where only the main sequence is represented, with subclasses B and A when the brightest stars in the cluster were B and A respectively. There was also a subclass F. In Class II there were giant stars in addition to the main sequence; Trumpler was very careful to establish real membership by means of the radial velocities. There was a third group, very rare, which again contained giants, but the number of giants was higher than the number of stars at the top of the main sequence. Today we understand these differences from the evolutionary point of view. Trumpler's material is a very valuable guide if one is looking for a certain type of cluster; his classifications make it simple to pick out clusters of that type. We shall mention some of his statistical results later.

Real progress came after the war, when it became possible to obtain color-magnitude diagrams essentially by photoelectric methods. Small difference between color-magnitude diagrams, which give us evolutionary information, actually need this very high accuracy.

Another important advance was the development, originally by Wilhelm Becker, of the method of three-color photometry. At least for blue stars it is possible to measure the brightness simultaneously in three colors, which yield not only the color-magnitude diagram, but also the absorption — a very important thing. The method was developed by Harold Johnson and Morgan, and led to the well-known U,B,V system; practically all the modern data on the color-magnitude diagrams of galactic clusters are on this system, which makes intercomparison of the data easy. Moreover, Morgan and Johnson determined very carefully the relation between the U,B,V colors and the Morgan-Keenan spectral classification. So most of our data now rest on a very high-grade unified system.

Among the very nearest galactic clusters are the Hyades, the Coma cluster, the Pleiades, and Praesepe. They are especially important because they are so close that we can determine the membership with high accuracy from proper motions and radial velocities. Usually there comes a point, for a cluster in which proper-motion data are not available, where field stars are so mixed in that it is difficult to follow the course of the main sequence. Thanks to the work of Hertzsprung, we have proper motions in the Pleiades down to 15^m, and, since the distance modulus is about 5^m.5, this means that we can sort out members with high accuracy down to 10^m. The cases of the Hyades, Coma, and Praesepe are similar. Of course, in case you have only proper-motion data, you must keep in mind that the proper motion of a star may agree very well with that of the cluster, and it may still not be a member.

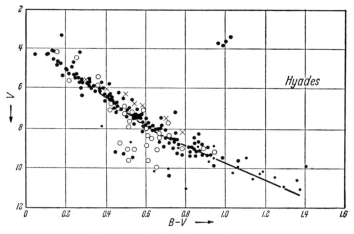

FIG. 7. Color-magnitude diagram for the Hyades.

The color-magnitude diagram for the Hyades (Fig. 7) is based on the photoelectric observations of Harold Johnson. A more complete and detailed diagram has been made by Johnson and Heckmann, partly based on the investigation of Praesepe by Heckmann and Haffner; it will have very high accuracy. The small number of stars at the faint end of the main sequence is not real, but a consequence of the small number of stars that have at present been meas-

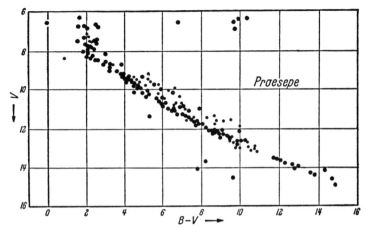

FIG. 8. Color-magnitude diagram for Praesepe.

ured photoelectrically. Four stars seem to be subdwarfs, lying on a line below the main sequence and parallel to it, but they have turned out not to be members of the Hyades. Greenstein and Münch checked the radial velocities, and found in each case that the radial velocity did not agree with that of the Hyades, so we can forget about these stars. They are simply stars with the same proper motion as the Hyades, which is not surprising, since the Hyades cover an area of 7° diameter in the sky, and there are lots of stars in this region moving in similar orbits. This shows how careful one has to be to sort out cluster members.

Stars shown by open circles in the diagram are possible members of the Hyades; they must all be investigated before we have a simon-pure color-magnitude diagram. I understand that the new diagram of Heckmann and Johnson, ranging from $+4^m$ to $+12^m$ in V magnitude, will take care of these stars. The diagram also contains four giants that have evolved from the main sequence.

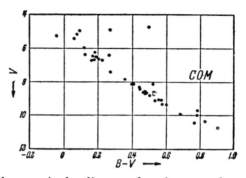

FIG. 9. Color-magnitude diagram for the star cluster in Coma Berenices.

The situation in Praesepe is very similar (Fig. 8): there are five G and F giants, and a well-defined lower limit to the main sequence. A number of stars lying above the main sequence were noted by Heckmann and Haffner as presumably binaries, and have been shown later to be spectroscopic binaries. It is likely that the other stars that lie above the main sequence are similar; their deviations are just of the right order. The fourteenth-magnitude stars that lie below the main sequence have not been checked yet, but this could be done today: very likely they are not subdwarfs, but simply field stars that happen to have the same proper motion as Praesepe.

The Coma cluster (Fig. 9) is of course a very sparse one. It shows essentially the same features: a well-defined main sequence and one or two giants; the observed range is 6^m. The proper motions are well determined. The Coma cluster illustrates the difficulties presented by galactic clusters as compared with globular clusters: the total number of stars is so small that there are very few turning off and evolving away from the main sequence — four in the Hyades, five in Praesepe, here perhaps one or two, com-

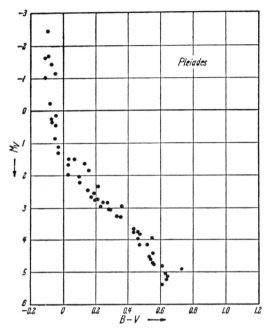

FIG. 10. Color–absolute-magnitude diagram for the Pleiades.

pared to the huge number found in globular clusters, simply because they are so rich in stars.

In the Pleiades (Fig. 10) we see that point at which stars are turning off from the main sequence, but there are no giants as yet. The faint end of the main sequence shows a very interesting thing: the stars plotted as black spots (the best candidates for membership) begin to fan out. This fanning out can be ascribed to the fact that some of these stars, at the faint end of the main sequence, are still contracting and moving toward it. So here we have the interesting case of a cluster, observed over a range of 11^m, which has an age such that the brightest and most massive stars are already evolving away from the main sequence, while the least massive stars are still converging to the main sequence. The fanning-out was observed at Leiden by Binnendijk who thought it might be an effect of lower accuracy either in the proper motions or in the magnitudes. But Johnson has shown that it is real. The data for the four clusters are given in Table 5. The

Table 5. Data for four nearby galactic clusters.

Cluster	$m_0 - M$	Distance (pc)	$M_{V,T}$	Number of white dwarfs	Age (yr)
Hyades*	3.0	40	+0.8	6	4×10^8
Coma	4.5	80	+0.5	?	3×10^8
Pleiades	5.5	126	−2.5	0	2×10^7
Praesepe	6.0	158	+0.8	several	4×10^8

distances have been determined anew in a final way by Harold Johnson, as I shall describe later. The true distance modulus, corrected for absorption, is $m_0 - M$; $M_{V,T}$ is the absolute magnitude of the brightest stars, the terminal point of the main sequence. Nothing is known of white dwarfs in the Coma cluster, but the Pleiades have been searched and none have been found. The Hyades are marked with an asterisk, because they provide the real zero point for the whole group: their distance is a geometric distance, based on proper motions and radial velocities, which has been redetermined for the new proper motions by Heckmann.

The age is inferred from the terminal magnitude of the main sequence. The interesting point is that two well-observed clusters (Hyades and Praesepe) that contain giants are close enough for the detection of white dwarfs, which start at 12^M, and run down, with increasing color, probably to 15^M. Six have been found in the Hyades, and four, I think, in Praesepe. In contrast, in the Pleiades, which contain no giants so far, not a single white dwarf has been found, in spite of searching. This is very interesting, because we have no idea of the evolutionary path by which the observed giants in the Hyades and Praesepe would become white dwarfs; there are no theoretical computations. However, we know that in about 4×10^8 yr (about the same for both clusters) some stars have already completed their whole life history from the main sequence, through the giant stage, and down to the white-dwarf stage.

The case is especially interesting in connection with Sirius B, which is a white dwarf, whereas Sirius A is an A star of absolute magnitude about $+1$. It is quite clear that Sirius B must once have been much brighter than Sirius A, and passed into the white-dwarf stage because it was more massive. Sirius A has an absolute magnitude corresponding to the upper break-off points of the Hyades and Praesepe. Thus the time taken for Sirius B to make its run was roughly 4×10^8 yr; of course if it was much brighter and more massive than the stars at the top of the Hyades main sequence, the time could have been much shorter.

On the way from the giant stage to the white-dwarf, most of the Population I stars must lose mass. If they did not lose mass fast enough, they would arrive at low absolute magnitude with an excess of mass, and probably would have to go through the characteristic nova-like processes. And the fact is that in Population I we do not find cataclysmic processes of this kind, though they should occur quite frequently. We can be sure that if stars reached the white-dwarf stage

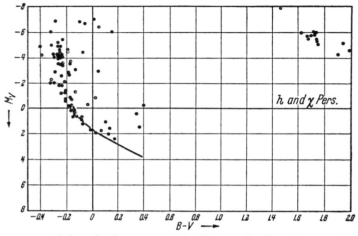

FIG. 11. Color–absolute-magnitude diagram for h and χ Persei.

with an excess of mass there would be cataclysmic processes, which are simply not known in Population I, except for the Type II supernovae, which are exceedingly rare.

We conclude that a Population I star with about 2 solar masses must arrive at the white-dwarf stage with a mass not much larger than 1.2 solar masses. There must be a mass loss that is not cataclysmic. The first clear-cut evidence of mass loss — a streaming out of matter during the giant stage — was obtained by Deutsch from observations of α Herculis. This is the first indication that mass loss is connected essentially with evolution during the giant stage. This means that we can apply the mass-luminosity relation with a fair degree of certainty as long as a star is on or near the main sequence. But by the time we reach the giant stage, we must be very careful, because there mass loss apparently takes place. I believe that the fact that we do not observe cataclysmic processes in Population I is a very strong argument for a steady mass loss at some phase, which

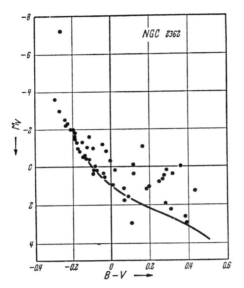

FIG. 12. Color–absolute-magnitude diagram for NGC 2362.

provides assurance that a star arriving at the white-dwarf stage does not need to get rid of any excess mass.

The clusters h and χ Persei (Fig. 11) contain stars a little brighter than -7^M. The large scatter is simply due to the fact that the region is badly cut up by absorption; one has practically to treat each star individually. Thus one cannot use the $U - B, B - V$ diagram in this case, as one normally does for a constant absorption. The color indices have to be determined with the aid of accurate spectral types. There is a group of M0 to M4 supergiants at $-5^M.5$.

The central star of NGC 2362 (Fig. 12) is the O9 star τ CMa. The main sequence is well determined up to the point at which it starts moving off. Here again we have a cluster in which some of the stars have settled on the main sequence, while some are still contracting toward the

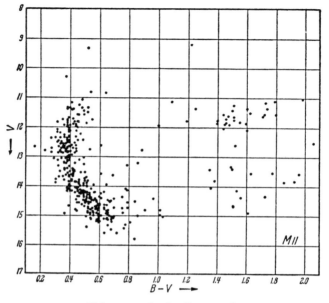

FIG. 13. Color-magnitude diagram for M 11.

sequence. The cluster is very young; from the figure —
we find an age of less than 1×10^6 yr; from the fact that
other stars are already on the main sequence we find an age
of 2×10^6 yr. The difference is explained if we assume that
the formation of the cluster occupied about 2×10^6 yr.

In M 11 (Fig. 13) there is again a considerable number of
giants. The diagram farther down is complicated by field
stars everywhere, because cluster members cannot be distinguished by proper motions.

In NGC 752 (Fig. 14) the brighter stars are turning away
from the main sequence toward the giant branch. It is very
difficult to fit the upper part of the main sequence onto the
standard sequence.

A very different cluster is M 67 (Fig. 15); the turning

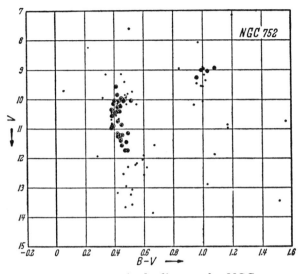

FIG. 14. Color-magnitude diagram for NGC 752.

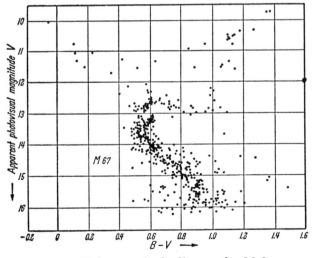

FIG. 15. Color-magnitude diagram for M 67.

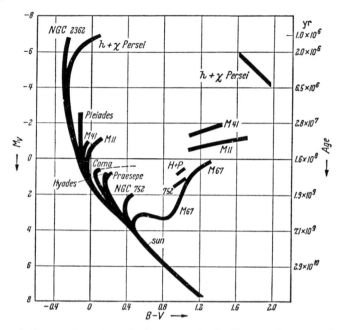

FIG. 16. Composite color–absolute-magnitude diagram for ten galactic clusters.

point is at about $+3^M.5$, as in a globular cluster, though it is not so well defined on account of the smaller number of stars.

We now refer all these clusters to a homogeneous distance scale (Fig. 16). For the Hyades we have a geometric distance, and in this way the main sequence of the Hyades can be calibrated in absolute magnitudes. The next step is to superimpose on the Hyades the main sequence of the Pleiades, which extends the diagram upward beyond o^M. Next we fit the diagram of h and χ Persei, and finally, at the top, that of NGC 2362.

The $U - B, B - V$ diagrams for all these clusters are identical. Johnson also intended to use the Coma cluster, but compared to the Hyades, Pleiades, and Praesepe this cluster showed a slight excess of intrinsic blueness, which suggests an effect of chemical composition such as that shown by the weak-line stars studied by Miss Roman. But the other three are exceedingly homogeneous on the basis of the two-color diagram. So in this way Johnson was able to establish a main sequence up to at least -2^M (not to the top of NGC 2362, for the brightest stars in that cluster are already turning off), and this can be used as a standard main sequence. For the faintest part he used the Hyades alone, not the Pleiades.

If we plot the stars in our neighborhood against this standard curve, we find that stars fainter than $3^M.5$ lie very symmetrically about it, whereas stars brighter than this lie, with few exceptions, a little above the main sequence: they have evolved far enough to have moved away. This is exactly what we should expect. It was necessary to change the relation to the Morgan-Keenan system by only $0^m.1$ in the critical part.

The main sequence forms an outer envelope in the composite diagram of Sandage (Fig. 16) that can be safely traced up to about -3^M; for brighter stars the evolution is so fast that we cannot establish it. As the absorption has been determined for each cluster, we can now relate them to one another. They are arranged according to their ages: NGC 2362, at the top, is the least evolved from the main sequence; its brightest stars are about the same as in h and χ Persei, which is much further evolved (age about 10^6 yr). Then come the Pleiades (about 10^7 yr), M 41, M 11, Coma, Hyades, Praesepe. One of the older clusters is NGC 752 (10^9

yr), and finally we have M 67 (5×10^9 yr), practically as old as the globular clusters. You see that the galactic clusters represent a whole series of ages.

In an old cluster like M 67, where the main sequence begins only at $+3^M.5$ and where many stars have already evolved, we should of course expect a large number of white dwarfs. As the distance modulus of M 67 is $9^m.6$, and white dwarfs usually set in at $+10^M$, they would appear at $19^m.6$. Two years ago I made a search of the central part of M 67 with the 200-inch. Since I had simply to go by color indices, I admitted only stars with color index smaller than $0^m.2$. I used only the bluest part of the white-dwarf series, because the series runs from very blue to very red stars — nobody knows at present how far — at the same time decreasing in brightness. About three dozen of these white dwarfs were found, between about 20^m to the limit of the plates at about $22^m.7$. From investigations by Greenstein and others, one would expect that about half the white dwarfs will be bluer than $0^m.2$ in a cluster of this age, so the number should probably be doubled, giving say 60 or 70 white dwarfs. When you consider that only the central part of M 67 was searched, the number may have to be doubled again, so that it is not impossible that M 67 contains between 150 and 200 white dwarfs. They will of course be quite difficult to investigate spectroscopically because the brightest are only of the twentieth magnitude.

At the same time I checked all the clusters in which we could hope to find white dwarfs — essentially, with our present equipment, Hyades, Coma, Pleiades, Praesepe and M 67. Coma is very difficult because it is so sparse and extended, but there may be quite a number of white dwarfs there.

What is the mass distribution, the mass spectrum, of

these open clusters? There are clusters where the main sequence terminates at $-7^M.5$, like h and χ Persei and NGC 2362; for others, like the Pleiades, it terminates at $-2^M.5$, and in others, like the Hyades, only at about 0^M. Of course the distribution can be anything in the Hyades, where stars are turning off the main sequence into the giant branch. But it is quite clear that there must be cases where the main sequence starts only with A stars. The lists of Trumpler and Stock contain quite a number of clusters whose brightest stars are about A2V, that contain no giants. All the stars from A2 downward lie on the main sequence, so there has been no evolution, which means that in these clusters the highest mass is about 2 solar masses. There are no very bright stars that can account for the missing part.

Even more interesting is NGC 2362, which is projected against a clear uniform field several square degrees in extent in Monoceros-Vela. There is no indication of any absorption, which is confirmed by the fact that the central star, τ CMa, is the bluest star known today. The general background sets in at or magnitude 14 or 15, but the cluster stands out far above the background, and seems to stop at about 12^m. This would indicate that in this case we have stars from O9, about 25 to 30 solar masses, down to about 2 solar masses, and there the thing stops.

So it is wrong to believe that these clusters are very uniform. The highest masses can have all sorts of values; we know unevolved clusters whose brightest stars are from $+2^M$ to -7^M, and on the other hand we have NGC 2362 which starts at about 30 solar masses and has nothing below 2 solar masses. This should emphasize that star clusters are a very peculiar phenomenon, and to say that they are just the same as the stars of the general field is certainly wrong.

There are two possibilities in determining the age of a cluster. First, we can use the point at which it is evolving away from the main sequence after burning its hydrogen. Second, we can use the computations of Henyey and his collaborators to determine the age from the stars that are still converging to the main sequence. In NGC 2362 both methods can be applied. From the fact that stars of about $+1^M$ are already on the main sequence, it follows from Henyey's calculations that the cluster is 2×10^6 yr old. On the other hand, from the fact that τ CMa is already moving away from the main sequence, we get an age less than 1×10^6 yr.

Now a great rumpus has been made about contradictions of this sort, but I think there is no reason to make any rumpus whatsoever. The two ages would be in strict agreement if all the stars were formed at the same moment; you can reconcile them without difficulty if you assume that it took about 2×10^6 yr, or even a little bit more, to reach the main sequence. The fact that some stars are on the main sequence and some are not is simply due to the time sequence of events: it took time for the stars to be created. It so happened that τ CMa was one of the early ones, and is already moving off.

Now consider the Pleiades. Here the age deduced from the thermonuclear process is 2×10^7 yr. According to the computations of Henyey and his collaborators, that would mean that stars fainter than $+4^M$, or $+10^m$, should still be in the stage of Kelvin contraction. But although, on this argument, the main sequence should begin to fan out at $+10^m$, it obviously begins to do so only at $+13^m$. Here we have the serious contradiction that the observed contraction time is very much smaller than the computed contraction time. The same contradiction, in the same sense, appears for the Hyades; I think it is a very striking thing. When we discuss associations we shall see that there the contradiction can be easily explained.

CHAPTER 10

STELLAR ASSOCIATIONS

O F   T H E   clusters discussed in the last chapter, only two are associated with nebulosity: NGC 2362, which is surrounded by a huge H II region over 200 pc across, and the Pleiades, which are surrounded by nebulosities. No nebulosity is at present known surrounding h and χ Persei.

We now turn to three objects of a very different kind; they are known as clusters at the present time, but perhaps we should classify them as associations. These are NGC 2264, which is associated with S Monocerotis; NGC 6530, which is projected against the well-known bright nebulosity M 8; and the stars associated with the Orion Nebula.

That these three objects are remarkable regions of the sky was found out very early. At Harvard an extraordinarily large number of variable stars was found surrounding the Orion Nebula at the same time as the variable stars in the Magellanic Clouds and in the globular clusters were being investigated there. Pickering realized that most of the variables in the Orion Nebula were probably irregular; from time to time people believed that they had found eclipsing stars or Cepheid variables among them, but these discoveries were not verified. Among the Orion stars there are roughly three types of variability; I worked on it myself in the 1920's. The most common type is a continuous irregular variation with amplitude of 1 to 1.5 magnitudes. Two others are relatively rare: in one type the star is usually of constant brightness and occasionally undergoes short or longer dips, and then returns to more or less constant brightness. The third type is probably still rarer: the stars are not visible at all for years, and suddenly shoot up from invisibility, may stay bright for a time, and then go down to invisibility again.

Around 1920, Max Wolf found another case of the same sort, the variable stars near S Mon and the surrounding nebulosities. In general the nebulosities are weak, so one can study the variable stars quite well. The remarkable thing was that every star brighter than a certain magnitude, in a very small area about 30′ × 15′, was variable or suspected to be variable. The association of the nebulosity with these variable stars was quite striking: outside the limits of the nebulosity the variable stars suddenly stopped.

A third example was M 8, where Lampland made a search in the 1920's and found large numbers of variable stars of this irregular type. In each case the NGC gave the description: "a star cluster surrounded by nebulosities"; the visual observers realized that star cluster and nebulosity were associated.

It was most tempting to study the Orion Nebula, because it is associated with the richest region of these variable stars; several hundred must be known today. But the investigation of the variable stars is exceedingly difficult because of the strength of the emission spectrum. The stars are faint, and they are most condensed in the dense nebulosity; the main absorption lines in the spectra of the stars are covered up by the big emission lines of the nebula. Investigations have been made in the outer regions of the Orion Nebula in order to circumvent this difficulty. Unfortunately the technical difficulties have prevented the investigation of the variables and of their association with the Nebula. Perhaps the suggestion, made by Herbig, of working far in the infrared may solve the problem, though according to some attempts made at Mount Wilson I am rather doubtful.

In the region of the Orion Nebula there is dust that scatters the light to a very, very high degree. A picture of the Orion Nebula on a U plate ($\lambda\lambda$ 6800 to 7200) reduces the effect of the nebulosity very greatly, though some very weak emission lines will still be included, but there is a strong region of scattered light, and it is hopeless to work in this area. The photograph shows the great increase of stars toward the center of the nebula, the region of the Trapezium. In the corner of the photograph is the normal foreground field containing very few stars. That so many stars are visible in the center is due to the fact that the Trapezium (in which the earliest spectrum is O, and there are two Bo stars) ionizes the whole region, and apparently has an effect on the dust there. In some way the dust must be modified by the photo-processes, so that at least in the infrared you can look into the nebula. In the blue everything is covered with nebulosities and in the unexcited region the cloud is completely dark. So the luminous part of the Orion Nebula is actually only part of a very much larger dark cloud. Short exposures on the Trapezium on infrared plates show nebulosity that is scattered light, not emission. Of the B stars around the Trapezium not a single one is of constant brightness. It would be very interesting to know whether the O star is variable.

A very much simpler region is that of NGC 2264, which surrounds S Mon. The star field all around is very rich, and there is a very impressive dark cloud, huge and elongated, running north-south. Herbig has found a large number of emission line stars with bright Hα in the region, and there is a remarkable association between these objects in the color-magnitude diagram. They are all restricted to the dark area. There is a very strong concentration to the center.

A nebulosity of the reflection type is associated with a B6 or B7 star, so reflection would be expected. There is another reflection nebula, and one mildly excited emission nebulosity, which shows only the lines of [O II] and H. That there is weak line emission over the whole area is shown by the fact that on all the slit spectra of the Hα region one sees very strong λ 3727 [O II], and also sometimes all the Balmer lines in emission. The Schmidt Palomar plates actually show that the whole region is enveloped in nebulosity.

Outside the region very few similar objects are found, although a careful search has been made, and the outer area would be more favorable for finding them because the dark nebulosity blots out all the faint stars, whereas outside it the spectrum of even the faintest star can be studied without difficulty. The central region falls into two definite areas — one around S Mon and the other around the B2 star that excites the enormous nebulosity — separated by a minimum.

It is clear at once that there is a connection between the emission-line stars and the dark cloud, which is consistent with the fact that most of the emission-line objects are T Tauri stars, which are only found associated with nebulosities. Obviously the dark cloud is quite near to us, because there are hardly any field stars projected against it, though outside it the field density is high. I think that the bright nebulosities and the dark one are parts of the same thing: we see the outer skin of the dark nebulosity. The bright-line stars are in this outer skin. Herbig has given a neat piece of evidence as to how far we can actually see into the dark nebulosity. He made star counts over the edge of the nebulosity to find out over what interval the stars disappeared, because the rest of the nebula is completely opaque. If we assume the distance, determined from the color-magnitude diagram, to be known, the stars disappear over distances of the order of 0.1 or 0.2 pc. This, of course, is only true for the edge of the nebula; the situation for direct incidence would be very different; at the edge we have certainly a strong gradient of density toward the center. Herbig made two assumptions of extreme cases: linear increase toward the center, and exponential increase, treating the whole thing as a cylinder. It turned out that for direct incidence one could not penetrate more than 1 to 2 pc into the nebulosity — of course an exceedingly rough result.

When Walker made a color-magnitude diagram he found four stars of relatively early type that showed severe reddening, and the number is quite reasonable. In contrast, the other stars outside this thin zone show practically no reddening. That is a very nice indication that the situation is something of the sort I have described. So we have a remarkably dark cloud, very dense; at the present time there are remnants of nebulosities excited by suitable stars and remarkably free from dust, gas, or absorption; and the stars that are deeper in the cloud are severely reddened. This fits in with what we found in extragalactic nebulae: the inner arms are dust arms, and along the edges, and occasionally outside, we see a few Cepheids. Of course we can only see the brightest stars, a sprinkling of supergiants and an occasional Cepheid. Here the cloud is so opaque that, although star formation is going on inside it, we see nothing of it. Part of the cloud is clear, probably due to the presence of one O star and some other early-type stars. But we know nothing of what is going on inside the cloud; most likely stars of high mass are being formed at a great rate. It would need the formation of a star of high luminosity to blow it up.

Such cases have actually been found. One of the best is IC 5146, which Max Wolf originally called the "Höhle" Nebula, or Cave Nebula; it is a huge bowl with an emission spectrum, all cut up by absorption patches, and in the center an O8 star. That is what you observe in the photographic region; you see very few stars because of the density of the nebulosity. But if you take an infrared plate, what do you see? If you take a U plate, which avoids the emission lines, you see the O star surrounded by the most beautiful cluster of stars. So in this case you can see through to the cluster of stars in the long wavelengths; it is a most striking thing. The case of IC 5146 induced me to check others.

So there is good reason to believe that if the cloud is not too thick, and if a big O star should be formed inside it, the whole region would be so modified that we could see into it. An H II region would form, and probably pass outward, and we should see at least a kind of hole. That is exactly what has happened with the Trapezium in the Orion Nebula; the Trapezium has blown a clear hole in the very much larger cloud, and the hole would be clearer if

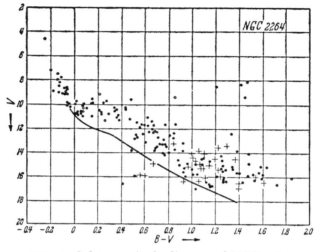

FIG. 17. Color-magnitude diagram of NGC 2264.

it were not for the continuum produced by scattering. The picture is complicated by a band of absorbing nebulosity, which is obviously nearer to us than the luminous Orion nebulosity. But the whole thing begins to make sense.

When Walker investigated the color-magnitude diagram of the stars surrounding S Mon, he restricted himself to stars lying within the confines of the dark nebulosity. In the dense field outside there would have been too many overlappings. The stars he measured are distributed over rather a large area, $1°.6 \times 0°.6$ which corresponds, with the final scale, to an area of 24×9 pc. I mention this because the diameters of normal clusters lie between about 1.5 and 15 pc; no known cluster has a diameter larger than 20 pc. So we are coming closer to associations, or these objects may already be associations and not clusters.

The plot of $U - B$ against $B - V$ shows that, so far as stars earlier than Ao are concerned, the absorption is very small.

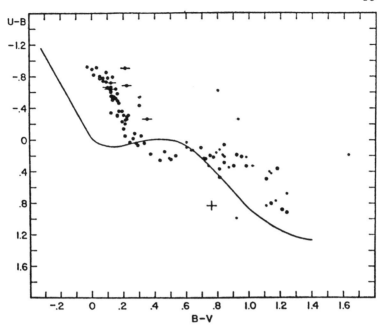

FIG. 18. Two-color diagram of NGC 6530. A solid line shows the normal curve.

The shift needed in $B — V$, which gives the reddening, is only of the order of $0^m.24$. The early stars, out to a color index of $0^m.7$ or $0^m.8$, fit the normal curve quite well. If you shift the curve over, this group of stars lies slightly above the horizontal branch. On the color-magnitude diagram (Fig. 17) you see that the brightest star of class O and stars down to color index $0^m.0$ (Ao stars) lie on the zero-age line. From then on the stars drift off the curve very rapidly. Stars with Hα in emission and variable stars are indicated in the diagram; all the variables with emission lines turn out to belong to the T Tauri class. In this cluster they set in at about 13^m to 14^m, which corresponds to absolute magnitude $+3^M$ or $+4^M$ since the distance modulus is about $9^m.7$.

The young cluster NGC 6530 is very similar to NGC 2264. The two-color plot (Fig. 18) is very interesting: the absorption itself is very small, but all the T Tauri stars lie far above the normal curve. The simplest explanation is that all these stars have large ultraviolet excesses, a striking indication that they are abnormal, although their spectra can be classified without much difficulty.

In the color-magnitude diagram of NGC 2264 there are five stars in the giant region. Walker has found that their spectra run from late G through K_0, K_2, K_3. We should interpret them not as normal giants, but as stars on the march to the main sequence. I think we should consider that star formation is going on in the dark nebulosity; it will probably continue for some time, because there is so much mass available inside the nebulosity. The observations of the outer skin, where we can see star formation today, suggests that these stars are still moving toward the main sequence; they may be following curves like the ones computed by Henyey. The stars that have just reached the main sequence are Ao stars of about 2 solar masses, which (with Henyey's data) gives an age of 2 or 3×10^6 yr.

The stars below the main sequence ($B - V$ about $0^m.6$) are very remarkable, and they are not understood at the present time. Since they resemble the T Tauri stars in every way, we cannot doubt that they are real members of the cluster. There will be a small number of foreground stars, but too few to affect the total picture. I think we must interpret these stars in evolutionary terms: in the dark cloud and on the outside, star formation started some time ago, and the stars are now moving toward the main sequence. In regions like the Orion Nebula the thing is very much more complicated; star formation set in earlier in the outer regions, and we can observe quite a number of areas in which star formation started at slightly different times. Certainly we cannot assume that all the stars in a region were formed at the same time; obviously they were not.

In analyzing such a cluster one must be very careful. I think we have reason to believe that there are no large irregularities in the absorption. The early-type stars on the main sequence, which lie in the edges of the nebulosity, cover the whole area, and they lie very closely on the main sequence, as we should expect. Since the other stars are intermingled in the same area, we can conclude that they are not affected by absorption either. The next step, which Walker has already started to carry out, is to determine the radial velocities for as many of these stars as possible; in this way we can go over the whole thing again with a fine-tooth comb and eliminate any intruders.

We know today that the T Tauri stars are intimately connected with dark and bright nebulosities, and so we cannot doubt that the group of stars below the main sequence is real, although we cannot explain it today.

The brightest star in the group is an O star ($B - V = -0^m.2$, $V = +4^m.5$). Now O stars are spectacular things; when one is present, your attention is drawn to it by nebulosities. Below the O star is a whole series of stars that seem to be normal, and would not be distinguished from the background in any way. Only when you reach a certain magnitude do you suddenly find the T Tauri stars, which can be detected either by their variability or by their emission lines.

A distinction is often made between O and B star associations and T Tauri associations. These are regions in which we can easily see that something is going on, but in the intermediate region it is not so easy to see. When Ambartsumian introduced the concept of associations he believed that only these two groups were involved, since there was no evidence of an intermediate group. He thought that stars are fed into the main sequence at two points — in the T Tauri region and at the top. I do not think that we believe this any longer.

For the stars whose spectral types have been determined, there is a correlation between spectrum and color index, both for the dwarf stars and for the giants, as there should be. There is no difficulty in classifying the spectra of these young stars, although they lie well above the main sequence. And as one goes to the right of the diagram, the spectral type advances with increasing color index. The spectral type runs from Ao through F6; the giants have spectra G_5, K_2 III, K_3, K_5. This cluster forms a link with NGC 2362, which includes τ CMa, but in NGC 2362 the series stops essentially at Ao, and there are no fainter stars.

We saw in the last chapter that the theories of gravitational contraction have difficulties with the time scale. According to the computations, we should expect from the upper main sequence of the Hyades and the Pleiades that stars should still be moving toward the main sequence at the lower end. For the Hyades this should occur at $+5^M$ or $+6^M$, for the Pleiades at $+4^M$, whereas we know that in the Hyades everything is on the main sequence down to $+9^M$, and in the Pleiades we see the tail end of the approaching stars at $+9^M$. There is a discrepancy of three or four magnitudes. We have no reason to doubt that the turnoff at the upper end gives the right date, especially in the Hyades, where we have the giants. The discrepancy is so large that we must ascribe it to a wrong time scale, and I think it is a very serious one. In the S Monocerotis cluster we see the same phenomenon as in the Pleiades; at the tail end of the main sequence we find the T Tauri stars and flare stars. The cluster NGC 6530, near M 8, is very similar. But whereas there were no difficulties in studying the cluster round S Mon by photography, because the field is clear, Walker found, after a heroic attempt, that he could not use photographic methods for NGC 6530. The difficulty is the same that I mentioned in connection with galaxies: the moment that you have a strong background you are in serious trouble. Walker used photoelectric photometry, and avoided the densest regions, because he had to search carefully around each star for test regions to get his zero, which slowed up the whole procedure tremendously.

Walker's search for variables was the first in this region, and there must be many more than he discovered. The absorption is very much larger than in the S Mon region; we are on the outer fringe of a huge nebula. Most of the light that excites M 8 itself comes from a star which lies deep in the dark region. Although Minkowski and I have taken a number of spectra, we have been unable to determine its spectral type; it is very early, probably an O star. But the reddening must be more than $1^m.5$, so to get the spectral type one would probably have to go far into the infrared, and spectral classification in the infrared is very difficult for early-type stars. But in M 8 there is really an O star already.

The final color-magnitude diagram for NGC 6530, after correction for the large absorption, is similar to that of the S Mon cluster. The brightest star in the cluster is again an O star, the points lie on the main sequence down to about $B - V = 0^m.0$, and then they run off. Walker avoided variable stars and emission-line stars that are likely to be variable, because he was working photoelectrically. There is no difficulty in interpreting the cluster as one in which star formation has been spread over an interval. The age of this cluster is probably of the same order as that of the τ CMa cluster. It contains one F giant, and there is no good reason to reject this star; it could be moving off, or it could be moving to the main sequence; we cannot tell.

These clusters are quite different from the run-of-the-mill galactic clusters without nebulosities, clusters that are so old that, if there ever were nebulosities associated with them, that phase is long over. Also the dimensions of M 8 and NGC 2264 are so much larger than galactic clusters that we might call them associations. We know of associations, such as that around h and χ Persei, with diameters of almost 300 pc. So star formation can go on over a large area. Among the extragalactic nebulae, there are beautiful examples where we are sure that star formation is in progress over an area 3000 pc across.

It would be very important to get more data about the Orion association, which we are still regarding in too narrow a sense. Apparently we should extend our investigations over the whole constellation of Orion. It is evident that star formation is essentially over in most parts of the constellation; the most active center is the Orion Nebula itself. We should expect to find associations or groups of all ages, marking out the whole history of the region.

In the plane of the Galaxy, where we observe dust and gas and star formation, we have overlapping groups that we cannot separate. We should therefore concentrate on the cases where we see a blob of sufficient mass, projected in higher latitude. The two ideal cases are the Orion Nebula and the R CrA region. I mentioned the technical difficulties that partly rule out the Orion region, but the R CrA region is really ideal. It is only 100 pc away, and is projected at galactic latitude 18°; it lies right above the galactic center. I have taken a series of plates marching up from the galactic center; there is no nebulosity until suddenly at 18° you encounter this beautiful complex. It think it would be a most fruitful group to study, and as it is distant only 100 pc, compared to 400 or 500 pc for the Orion Nebula, one could really do something. There are some B stars in the region which probably excite the nebula. But this work must be done in the Southern Hemisphere, for the declination is about —40°. The region would also be an easy one for investigation in the radio range, because it is so nicely isolated that there should be no difficulty with overlapping of nearby objects.

We must keep in mind that we have only the first raw attempts on these groups of stars, and that there is no sense in overdiscussing them yet. Everything should be done to verify these precarious foundations, with the aid of radial velocities and everything else we can use, in order to get a residue of fact on which we can bank.

CHAPTER 11

THE T TAURI STARS

We have discussed the color-magnitude diagrams of clusters and of associations. The composite diagram for galactic clusters (Fig. 16) includes a series of all ages, from NGC 2362, one of the very youngest, to M 67, one of the oldest. The series is highly selected; of course it does not represent the real frequency of cluster ages. It would be exceedingly interesting to know what is the distribution of the ages of the clusters that we observe.

The age distribution will obviously be determined by two factors — the rate at which clusters are generated, and the rate at which they finally dissolve, become unrecognizable or invisible, and disappear as clusters. We should soon have pretty good information about the rate at which galactic clusters are generated in our neighborhood, because the youngest clusters are generally also the brightest. Harold Johnson is at present investigating a number of clusters, and good data should be available pretty soon.

I have made an attempt to see whether one could get an idea of the age distribution from the 100 clusters observed by Trumpler. But although Trumpler observed 100 clusters, they must have been largely selected according to apparent brightness — which is no wonder, because old clusters will be inconspicuous and small. The results, which are given in Table 6, have no significance. You see imme-

Table 6. Age distribution for 100 clusters.

Earliest spectrum (main sequence)	Age (yr)	Percent
O	$1- 2 \times 10^6$	10
B0–B7	$2–30 \times 10^6$	40
B8–A2	$3–50 \times 10^7$	20
A3–A7	$5–17 \times 10^8$	28
F0 and later	$>3 \times 10^9$	2

diately the undue preference for the younger clusters, especially in the very rapid dropping off around Fo, a consequence of the selection according to brightness. It remains to get the real distribution for a given volume of space, and there is no doubt that this will be very difficult.

It would be a very nasty job to get color-magnitude diagrams for all these clusters. Perhaps one could partly get around it by determining just the spectral types of the brightest stars by investigating the clusters with objective-prism plates. A recent attempt — apparently very successful — has been made by Stock with the 32-inch Schmidt at Hamburg and with the ADH Schmidt in South Africa. For faint clusters the Schmidt will not be quite sufficient, but I think that with an instrument like the very effective spectrograph used with the Crossley at the Lick Observatory it would be possible to reach them. I believe such observations will be necessary.

The list given by Stock contains quite a number of faint clusters, and the remarkable thing is how many have O and OB stars as their brightest stars. The list of all known color-magnitude diagrams put together by Miss Barkhatova contains practically all early types; it is astonishing how many there are.

It will be important to study the distribution in space of all the clusters that must be observed, many of them not especially interesting in their color-magnitude diagrams. We shall need to know the concentration to the plane of the Galaxy, the location in spiral arms, and how far they can finally move out of the spiral arms. Here the data are limited at present, but we should soon have very good information. The relation to the spiral arms will be exceedingly important, as a check on the location of the arms. The radio data depend on rotational theory, and the clusters are ideal, because their distances are so much better determined than that of any individual star.

Wilhelm Becker, who has worked rather continuously on open clusters in recent years, has discussed the distribution of 40 clusters that he has observed, 38 of which are clearly located in the nearby spiral arms. There are 28 in the Orion arm, which passes through the sun; 13 in the Perseus arm, the next outer arm; and 3 in the next inner arm, the Sagittarius arm. These data are very important in relation to both age effects and distribution in the Galaxy.

These star clusters fit rather perfectly into the evolutionary picture. The situation was different when we discussed the nearest associations. The association around S Mon is related to a dark cloud which cuts out the background entirely. If we make the justifiable assumption that the cloud is at the same distance as the association, its longest diameter is about 30 pc, and it is probably one of the smaller clouds in the spiral arms. In most of these associations everything agreed with evolutionary theory: the ages derived from the brightest stars still on the main sequence and from the stars that had just reached the main sequence were in pretty good agreement. For the S Mon cluster the stars still on the main sequence gave an age of 2.8×10^6 yr, or less; the age from contraction was 3.6×10^6 yr. The stars that were approaching the main sequence lay above it, but there were some stars below the main sequence that should not have been there. I pointed out in the last chapter that there seems to be a contradiction for the Pleiades: we know the age to be 2×10^7 yr and by the contraction theory stars down to $+3^M.7$, or $+9^m.2$, should be on the main sequence. Actually we know that the stars only begin to turn away at $+6^M.7$; above that, everything is on the main sequence. This is a crystal-clear case: the $+6^M.7$ corresponds to $+12^m.2$, and the Pleiades sequence is well known down to $+16^m$, so we have ample leeway. The predicted age is wrong by a factor of about 10. This must simply mean that there is some factor that makes things work very much faster, at least in this range of luminosities.

The two-color diagram $(U - B, B - V)$ for the S Mon cluster shows some features that seem to me important. The stars lie close to the main sequence — a one-parameter series. At the bottom of the main sequence the bunching is still pretty good but the scatter becomes larger, and suddenly, at $B - V = 0^m.7$, everything goes haywire. The culprits are the T Tauri variables, which simply do not form a one-parameter series; in the color-magnitude diagram they lie below the main sequence.

Among the stars of associations (such as S Mon, M 8, and Orion) we find a whole series of these remarkable stars, most of which are variable, and a high percentage of the variables have strong emission lines. Only the stars with strong emission lines are counted among the T Tauri variables. A star may not have emission lines for a long time, and may acquire them; or another star may lose them.

The spectra of T Tauri stars (according to Herbig, who has studied them more intensively than anybody else) begin at about F 8 and run up to about M 3; the luminosities are from about $+4^M$ to $+10^M$ and fainter.

The typical bright lines in the spectra of T Tauri stars are the H and K lines of Ca II, the hydrogen lines, and the lines of Fe I, Fe II, Ti I, and Ti II. But the remarkable thing is that in moderate T Tauri spectra the bright lines are not very intense; you can see the continuous spectrum without difficulty and can apparently classify them as easily as a normal spectrum. But when the emission lines are strong, a very remarkable thing happens. Not only does the number of faint emission lines become very large, but also (as Herbig found out at the coudé focus of the 200-inch) there is a continuous spectrum overlying the whole thing, so much so that it is impossible to see the line absorption spectrum. The continuum may occasionally clear, and one can at least get glimpses of the underlying absorption spectrum. The source of this continuum is still a matter of debate: Ambartsumian suggested that it originates in the synchrotron process, relativistic electrons moving in a magnetic field, but whether this is so will be difficult to decide.

In any case there is no doubt that the T Tauri stars are very remarkable objects.

These emission lines have an average width corresponding to a velocity of the order of 200 km/sec. Moreover, there are absorption components of shell type that are shifted to the violet by about 150 km/sec (exceedingly well determined by the coudé spectrograms), so there is every indication of outward motion of matter, whereas there is no evidence that matter is falling in.

I think that here we may have a clue to the trouble that the stars of lower masses appear too early near the main sequence. If there is a process by which they can lose a sufficient amount of matter in the latter stages of contraction, when they have already moved close to the main sequence, there may be a way out of the difficulty. Whether there is we do not yet know; it would be necessary to measure how much mass a T Tauri star loses.

About 40 to 45 percent of the variable stars in these groups are in the T Tauri stage. So far as I know there are no indications of mass loss among the other variables. This would mean that 40 percent of the variables go into the T Tauri stage and lose mass. Another interesting point is that the T Tauri stars, in contrast to the other variables (which have not, however, been as carefully investigated) show considerable line broadening. The data again are based on spectra taken at the coudé focus of the 200-inch. If you interpret the broadening as due to rotation, as Herbig proposed at one time, you get rotational velocities up to 50 or 60 km/sec. You may also interpret it as due to large-scale turbulence (which I prefer), which might be possible because one might expect a deep outer convection zone for stars in this stage.

Now comes a very remarkable thing: probably all T Tauri stars are surrounded by nebulosity. Haro was the first to pick up these objects — stars that are surrounded by a small nebulosity of emission type. Herbig found some more, and today they are known as Haro-Herbig objects. The star T Tauri itself is one; it has a very small nebulosity around it, as was discovered at the Lick refractor by Burnham, just a few seconds in diameter, and of nice elliptical form. With short exposures, the images are elliptical.

Herbig has taken the spectra of a large number of T Tauri variables, and in practically every one he finds [S II] lines, which are conveniently located at λλ 4068 and 4076. Occasionally he finds the [O II] lines at λ 3727, but the most persistent lines are those of the [S II] pair, which show up in practically all the T Tauri stars. Herbig had suspected for a long time that these [S II] lines were not connected with the star, because they simply did not fit in with the rest of the spectrum, but might belong to the surrounding nebulosity. When he turned to the investigation of the extended Herbig-Haro objects, he found indeed that the [S II] lines came from the whole of the nebulosity, as did the weaker [O II] lines and those of hydrogen. The moment he went from the star to the nebulosity the broad lines became quite narrow. So there was good reason to believe that all the T Tauri stars are surrounded by such nebulosities, but they would be very small and in most cases would pass unnoticed, since even for T Tauri, only a little over 100 pc away, the diameter is only a few seconds.

It is very interesting to inquire what excites the nebulosity surrounding the T Tauri stars, and the question was taken up by Wurm and Osterbrock. From the continuous spectrum the stars appear to be in the range from F8 to F3, too cool to excite these nebulosities. The problem was solved very neatly by Osterbrock. In addition to the [O II] lines, the nebulosities show the [O I] pair at λλ 6300, 6363. The first ionization potential of O is about the same as that of H, so the two atoms must be ionized to the same extent. Observation showed that the [O I] and [O II] lines are about equally strong, so that O, and consequently also H, is only partially ionized. Thus the material in the nebulosities is only partly ionized.

The next step was to find the density in the shell, which Osterbrock obtained from the intensity ratio of 0.56 ± 0.04 for the two components of the line at λ 3727. This led to an electron temperature of 7500°K and an electron density $N_e = 3700$ cm^{-3}. This was in good agreement with the values of 7500°K and 2500 cm^{-3} that had been derived by Wurm from a similar comparison of the intensities of the lines of [O II] and [O III]. The ionization was determined from the ratio of $O/O^+ = 1.9$, and thus there are two hydrogen atoms and one hydrogen ion for each free electron. With this figure, and the assumption that the H/He ratio is 2, as for the Orion Nebula, we obtain a mass density of 3×10^{-20} gm/cm^3 for the nebulosity. Since the radius is 1400 A.U., the total mass of the shell is quite small, 10^{30} gm $= 5 \times 10^{-4}$ solar masses.

If you try to excite a nebula of this sort by normal photo-ionization, you need a star of about 1 solar mass at a temperature of 20,000°K. But the spectrum of the nebula would of course be very different from what is observed: the nebula would be almost completely ionized, and only in the outermost regions would there be a rapid transition from almost complete ionization to neutral hydrogen. The lines of [O I] would be extremely weak, because they would originate from an exceedingly small zone, so the ratio of [O II] to [O I] would not be as observed. So the ionization is certainly not by this process. What then produces the luminosity of the shells surrounding the Herbig-Haro objects? The central star has too low a temperature. I think that Osterbrock's interpretation is probably right. We have seen that the star is ejecting matter with velocities of the order of 150 km/sec, which would correspond to electron energies of 100 ev. He suggests that this matter hits the shell and is responsible for the excitation of the nebulosities. In my opinion, the most interesting feature of the T Tauri stars is the mass loss that they show.

How long does the T Tauri stage last, when this ejection is going on? We have only one piece of information. We know that the T Tauri stars are intimately connected with dust and gas; we never see them against a clear part of the sky, only at the fringes of emission nebulosities. That means that during this stage they simply cannot have moved far from their place of origin. To take a very rough figure, we get an indication that the stage can last perhaps 10^6 yr, maybe 5×10^6 yr as an upper limit.

The next question to be settled is: how much mass does such a star lose in 10^6 to 5×10^6 yr? It is not impossible that the loss is great enough to throw some light on the puzzle presented by the stars' appearing already so close to the main sequence. I think that more information will come from observation, because among the dwarf stars associated with nebulosity and obviously in a state of formation there is another extremely interesting and exciting group, namely the flare or flash stars.

There are flare stars in our neighborhood, such as UV Ceti, which go up in a matter of minutes by 1^m or $1^m.5$, and decline just as rapidly. Such stars have also been found in large numbers in the association I have mentioned, the

Orion Nebula, S Mon, and M 8, and they behave essentially as flare stars do. Harold Johnson found the first flare stars in the Pleiades, which contain quite a number of them. The observers of the flare stars have laid stress on the fact that the spectral type differs from one group to the next. In the dark Taurus cloud we find them only at class dM3 and later. In the Orion Nebula, they are not earlier than K, and in S Mon, which is younger still, the first is already a G star. It will be interesting to find the spectrum of those in the Pleiades; it should be very late, probably M 7. So this stage is probably connected with the evolutionary pattern.

CHAPTER 12

GLOBULAR CLUSTERS:
COLOR-MAGNITUDE DIAGRAMS

IT IS quite difficult to establish standards down to magnitudes 21 and 22, even with the photocell. The observations are very time consuming, and therefore progress has been slow. The only globular cluster so far in which the color-magnitude diagram has been extended to the main sequence is M 3, by the work of Johnson and Sandage. Soon we shall probably see M 92, M 5, and M 13 extended to the main sequence.

The color-magnitude diagram of M 3 on the new scale is shown in Fig. 5; Fig. 6 compares the mean color-magnitude diagram of M 3 with the main sequence for the Hyades. The zero point of the absolute magnitudes for M 3 depends on an assumption about the absolute magnitudes of its cluster-type variables.

The Hyades-Pleiades sequence is final: it was determined geometrically. There is a difference of about $0^m.3$ between the main sequences of M 3 and the Hyades, and if we assume that the two main sequences should actually lie close together, despite differing chemical composition, there is thus a discrepancy. It could be removed if we shifted the photographic zero point of the cluster-type variables from $0^M.0$ to $+0^M.3$.

The Shapley zero point of $0^M.0$ has been adopted for the cluster-type variables; Shapley's value was of course on the International System. We should now determine the value of M_V for the cluster-type variables on the $B - V$ system. It turns out that the value is $-0^M.15$ instead of $0^M.0$.

In Chapter 8 I mentioned the determination of the zero point by means of four Cepheids that are members of galactic clusters. Arp and Sandage have determined the correction ΔM to the zero point on the International System to be $-1^m.27 \pm 0^m.14$. This correction applies to the classical Cepheids. We must now, to obtain the zero point for the cluster-type variables, apply the other correction, the difference between cluster-type variables and Cepheids, as determined from the Magellanic Clouds and otherwise. This difference is $-1^M.50$. So the cluster-type variables should be moved down by $+0^m.23$, which would lead to a value for M_V of $+0^M.08$. The main sequence of M 3 should therefore be shifted down by $0^m.08$. Of course this is still rather uncertain, but the accuracy of the Hyades-Pleiades sequence is high, and in the next few years we shall use more Cepheids and be able to rely on our zero points.

Although we still have to wait for extension of the diagrams down to the main sequence, we have Arp's investigation of the color-magnitude diagrams of seven globular clusters over the giant and subgiant branches. I shall divide them into three groups: (1) M 3 and M 5; (2) M 15, M 2, and M 92; (3) M 13 and M 10. The diagrams are shown in Fig. 19.

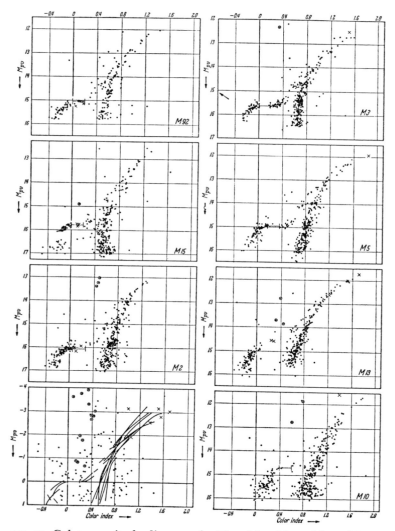

FIG. 19. Color-magnitude diagrams for M 3, M 5, M 13, M 10, M 92, M 15, and M 2. The lower left-hand figure is a composite of all seven of the globular clusters.

To agree on terminology: we call the part above o^M the giant branch, and the part that turns downward, the subgiant branch. The horizontal branch succeeds the turndown; it includes the cluster-type gap. In M 3 and M 5 there is a well-developed horizontal branch, and there are stars on both sides of the gap. In M 15 and M 2 the horizontal branch is weaker, but there are still cluster-type variables. In M 92 the horizontal branch is strong to the left of the gap, weak to the right. There are just two cluster-type variables in M 13, and in M 10 the gap is entirely fictitious, because there are no cluster-type variables at all.

If now one superimposes the diagrams, one can easily see, without driving the comparison to any finesse, a number of features. The diagrams of M 3 and M 5 can be brought into coincidence perfectly over the whole range. The characteristic feature of this group is the strong development on the red side of the gap. As we go to the next group, the horizontal branch on the red side of the gap begins to weaken. Moreover, relative to M 3 and M 5, the giant and subgiant branches are shifted over bodily toward the blue. The largest shift is shown by M 92, where it amounts to −0^m.17 in color index on the International System: the whole thing is shifted over bodily. The diagrams of M 13 and M 10 coincide essentially with those of M 3 and M 5 in the giant branch. But the subgiant branches in M 3 and M 5 are steep, whereas in M 13 and M 10 they run at a very much smaller angle. The divergence begins at about −1^M.0 visual, just about a magnitude above the cluster-type variables. The diagrams of other globular clusters all conform to one or other of these patterns.

We know pretty well, from the precision work on M 3,

how wide the cluster-type gap is in color; how wide is it in absolute magnitude? How close does a star have to be to certain conditions to be a cluster-type variable? There is a star in M 5 that lies perhaps $0^m.2$ above the gap; it would be exceedingly valuable to check this star.

The subgiant branch of M 15 comes down at a color index of $+0^m.55$; in M 92 it is close to $+0^m.45$. In M 13 we also have a shift, but for a different reason: the upper branch coincides with those of M 3 and M 5, but the slope continues instead of dropping down steeply.

In order to make a composite diagram for all the clusters, it has been assumed that the blue end of the variable-star gap is constant, both in color and in absolute magnitude. The red end becomes inconspicuous, and one could not fit it very well for all the clusters. We must keep in mind that the diagrams have been fitted in this way. The composite diagram for the clusters is shown in Fig. 19. In addition to the mean lines, a number of stars that do not lie on the mean lines have been included; they are certainly significant, for very blue stars should not occur at these high latitudes. In spite of minor differences, the general picture is the same for all the clusters; the topology is the same.

Table 7 shows the distance from the red end of the gap to the giant-subgiant branch, in an arbitrary measure, for these clusters. The value for M 15 is pretty close to that for the first group, but in M 2 and M 92 the giant branch is shifted toward the red. In M 13 and M 10 it has moved part of the way back again, so these clusters are not a continuation of the preceding group.

It is obvious right away that the density along the horizontal branch, and especially on the red side of the gap, is related to the frequency of the cluster-type variables. Clus-

Table 7. Distance (arbitrary units)
from red end of gap to giant-subgiant
branch in seven globular clusters.

Cluster	Distance
M 3	27.0
M 5	26.5
M 15	26.0
M 2	21.5
M 92	21.0
M 13	24.0
M 10	24.0

ters like M 3 and M 5 are rich in variables, but M 13 contains only two, and M 10 contains none. The second group are intermediate cases.

The globular clusters have been divided into two classes according to the frequency distribution of the periods of their variables. In both classes the short-period variables have sinusoidal light curves, and the variables of longer period have asymmetric curves. In Class I, the shorter periods have a mean period of about $0^d.33$, the longer periods, $0^d.525$; numerically, the former group is poor, the latter, rich. In Class II the mean periods are $0^d.37$ and $0^d.625$, and the two groups are about equally numerous.

It turns out that M 3 and M 5 belong to Group I, and M 15, M 2, and M 92 to Group II; M 13 and M 10 contain too few variables, so we cannot say anything. Remember that you can superimpose M 3 on M 5, whereas the next three are shifted bodily, with the greatest shift for M 92. Since the position of a star in the color-magnitude diagram is determined by mass and chemical composition, we can

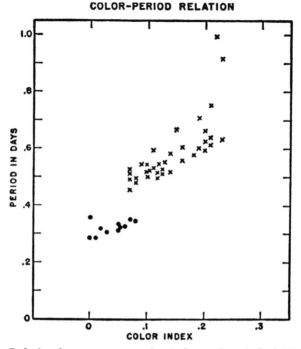

COLOR-PERIOD RELATION

FIG. 20. Relation between mean color index and period of RR Lyrae variables in M 3.

perfectly understand a bodily shift, if the stars are all built on the same model and have the same mass, but differ in chemical composition. In fact, Schwarzschild and Sandage have shown that the shift of color of o.^{m}17 can be explained if to M 3 and M 5 is ascribed a frequency of metals only 1.3 times as great as in M 92. What cannot be interpreted in this way is the difference in slope of the subgiant branches. A change of model might account for this, and it could be for reasons that we do not understand, or a change in what determines the opacity. These things can be settled only if a number of models are computed. It is quite possible that

PERIOD-AMPLITUDE RELATION

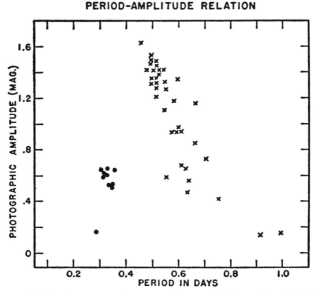

FIG. 21. Relation between photographic amplitude and period of RR Lyrae stars in M 3.

the remarkable division in the periods may be a first indication of the influence of composition on the properties of variable stars.

It is obvious from these diagrams that the appearance of cluster-type variables is intimately connected with the number of stars in the horizontal branch, especially on the redward side of the gap. The picture is that the stars move to the left through the gap, and then decline. An important piece of evidence has been obtained by Sandage for M 3. If you integrate the color index of a cluster-type variable over a cycle, and plot against period, you get a very close correlation indeed (Fig. 20). You can see where the stars from the red side of the gap are coming in and producing a correlation between color index and period. The relation does not contradict the idea that the variables are moving through the gap. There is also a correlation between period and light amplitude (Fig. 21).

Do we deduce reasonable changes in radius, surface brightness, density, and so on, from our data? I have tried to stay away from any data that involve bolometric magnitudes. To deduce the decrease of radius for a star as it moves across the gap, I have calculated it for the long-period and the short-period group of this type of variable. I find that the radius decreases by about 30 percent on the march through the gap. If the subscripts 1 and 2 refer to the beginning and the end of the gap, then $(R_2/R_1) \propto (P_2/P_1)^{2/3}$. The mean density triples as the star passes the gap: $(\rho_2/\rho_1) \propto (P_1/P_2)^2$. The surface brightness S doubles: $(S_2/S_1) \propto (P_1/P_2)^{4/3}$. Finally, what does the temperature do? Of course we have to use the Stefan-Boltzmann law, and the star probably does not radiate like a black body, but if we assume it does we find that, since $(T_2/T_1) \propto (P_1/P_2)^{4/3}$, the surface temperature increases by 20 percent. I think there can be no a priori objections to such changes; they are quite on the moderate side.

How wide is the gap, and has it the same width in all clusters? Again our information is very scanty at the present time; we have good measures of the gap only for M 3 and NGC 4147. In these two clusters the gap in color index is the same within $0^m.02$: for M 3, the width in color index is $0^m.24$ on the $B - V$ system, and for NGC 4147 it is $0^m.22$. About the extent of the gap in the vertical direction we know nothing at the present time.

I think that is about all one can say at the present time concerning the color-magnitude diagrams of the globular clusters. Topologically, as we have seen, they are all very similar. Publication of the data on M 3, M 13, M 5, and M

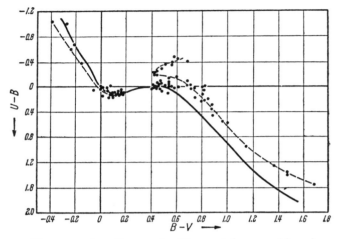

FIG. 22. The two-color diagram for M 3 and for unreddened nearby stars (solid line).

92 should give us new information on how far the turning-off points from the main sequence agree.

I should like now to discuss the $U - B, B - V$ diagrams of the globular clusters. The question is how far the stars in globular clusters fit the normal relation. Again the data are still quite scanty; we have good material now for only three clusters: M 3, M 13, and NGC 4147, which are all very favorably located in high galactic latitude. The best-observed cluster is M 3, and here it turns out that the stars of the horizontal branch all fit the curve very well (Fig. 22). But the stars of the giant branch, starting from the bottom of the diagram, lie above the curve and approximately parallel to it. Then comes the main sequence, which turns off and deviates widely. The globular clusters differ in a very systematic way from the two-color representation of what we call normal stars in our own neighborhood. These diagrams finally put the fear of God into those fellows who were glibly manipulating colors and bolometric magnitudes, for I think this diagram should convince anybody that it is just nonsense to operate with the normal relations between color and spectral type or color and temperature for globular clusters. These relations must be derived entirely anew from the data for clusters.

In M 13 another feature shows up — an ultraviolet deficiency for stars in the horizontal branch, which now lie too low, even after correction for any reasonable amount of absorption. With increasing color index, the stars again lie above the normal curve, and as you go still further the difference disappears to some extent. You must take into account, however, that the standard curve is for main-sequence stars, and the curve that we compare in the clusters is for the giants. The diagram for NGC 4147 is similar: the horizontal branch apparently fits rather closely, but the moment you come to the giants and subgiants the whole curve lies not only far above the curve for normal stars, but also above the curve that is observed among the high-velocity stars of our neighborhood. This means that these cluster stars simply do not resemble those of the galactic field, even over the limited range from the accessible ultraviolet to about 6500 A.

The difference was strikingly emphasized when the spectra of these cluster stars were taken a few years ago with high dispersion at the 200-inch. The stars observed were essentially those at the top of the giant branch, with color indices from +1.3 to +1.6, and from the normal relation between color and spectrum one would have expected late K stars. But the actual spectra were a great surprise, because the classification depended on what lines were used. If one used the metallic lines, the spectral type (especially for M 92) was consistently around F3, whereas from the colors one would have expected K3. The hydrogen lines gave something like G3. The discrepancy was milder for M 3, where the expected spectra were about K7, and the best one could do from the hydrogen lines was something near Ko. Joy and Popper had studied these spectra with smaller dispersion, so that they had to rely essentially on the hydrogen lines, the G band, and so on, and had obtained similar results, showing up, in one stroke, how far the stars in globular clusters deviate from the relations that we know for nearby stars.

This striking difference must be interpreted as a result of a real difference in chemical composition, in the sense that these globular-cluster stars are very deficient in metals, compared to stars in our neighborhood. A few years ago, Hoyle and Schwarzschild undertook to compute the evolutionary path of a star that leaves the main sequence at absolute magnitude +3.5, has a mass of about 1.5 suns, and has a metal-to-hydrogen ratio 1/17 that for stars in the solar neighborhood, and succeeded in representing the general form of the giant branch in globular clusters. There they had to stop, but up to that point the representation is indeed good. They also computed the evolutionary path for a star in which the metal-to-hydrogen ratio is the same as for the sun, and found that in that case the curve would run below that for metal-poor stars, and the giant branch would reach a maximum luminosity around zero, whereas for metal-poor stars the curve ran up to absolute magnitude −3. These results appeared just when Sandage had determined the color-magnitude diagram for M 67, which showed a turnoff point like that of the globular clusters. As M 67 fitted quite well with the calculations for stars of normal metal content, it was evident that the difference between M 67 and the globular clusters must be one of chemical composition. It now became clear that the giant stars in M 67 cannot have absolute magnitudes as high as —3 because the opacity is provided in this case by the electrons furnished by the metals. The difference in spectral classification, then, makes sense: the lines of the metals would be very much weaker for the cluster stars, and the spectral type indicated by the metal lines would therefore be very much earlier than that indicated by the hydrogen lines.

Since the hydrogen lines give essentially the excitation temperature, we conclude that the excitation temperature corresponds to a spectral class one whole class earlier than we should expect from the color index. Or you can put it that the lines of the metals give a spectrum that is a whole class too early, and the color index leads to a spectrum that is a whole class too late; there you have the whole difficulty, which must be disentangled by means of new computations. In short, the Population II stars in globular clusters are so different from those in our neighborhood that the relations between color index, temperature, and spectral type will have to be determined all over again for them. Computations of evolutionary paths have now been started in order to take intermediate cases into account, and also to extend on both sides to lower and higher metal-to-hydrogen ratios, and these computations are very important, because data that are now coming in might otherwise be misinterpreted.

On the basis of the spectra of stars in a number of globular clusters, and of the color-magnitude diagrams, we find already that there are clearly recognizable differences among the globular clusters of the halo. One of the next jobs will be to make a real quantitative analysis of a few stars in some of these clusters, and I am afraid it will be frightfully complicated. For instance, we need to know the relation between the temperature and the distribution in the continuum, because it is obviously very dangerous to work with the standard relations, and can lead to completely erroneous results.

CHAPTER 13

SPECTRA OF

POPULATION II STARS

I N T H E last chapter I mentioned the spectral peculiarities of the stars in globular clusters — the discrepancy between color index and spectral type, and the differences between the $U - B, B - V$ diagrams for clusters and normal stars. I pointed out that the thing makes sense in the light of the computations of Hoyle and Schwarzschild: the globular cluster stars have a true metal deficiency.

Since the weakness of the metallic lines differs among the globular clusters of the halo, it follows that there must be differences of chemical composition among them. In M 3 the giant and subgiant branches are most normal, and come closest to standard stars, and the largest deviation is for M 92.

Another easily-observed characteristic of the spectra of stars in globular clusters is the weakness of the cyanogen bands at 4215, 3800, and 3550 A. Lindblad noticed this effect for the first time as long ago as 1922. He determined the spectral types of three of the stars in M 13 on a slitless spectrum taken at Mount Wilson, and noticed that these three giants had abnormally weak CN bands. For this reason he called them "pseudo-Cepheids," because he believed the the CN bands decrease in intensity from giants to supergiants, but Keenan showed later that in this spectral range (Go to about K3) the intensity of the CN bands increases through the subgiants, giants, and supergiants.

That the weakness of the CN bands is a characteristic feature in globular-cluster spectra was shown more extensively by Popper in 1946, when he studied about a dozen stars in each of the globular clusters M 3 and M 13. In principle, this result depends on the low abundance of C and N. The spectral types obtained by Popper for the five brightest stars in M 3 ranged from G5 Ib to Ko Ib, the mean being G8 Ib; these types were essentially based on the hydrogen lines. The mean color index for the same stars was $+1.65$ on the new $B - V$ system, and the mean absolute magnitude was -3.1. What spectral type should we expect for these stars? In this range of spectrum, the relation between spectral type and color index is known at present only for main-sequence and giant stars; for supergiants we have to extrapolate. However, the supergiants in h and χ Persei, which have been carefully investigated by Sharpless, provide a very fine fixed point. These stars have a mean absolute visual magnitude of -5.5, an intrinsic color index (freed from absorption) of $+1^{\mathrm{m}}.68$, and a mean spectral type of M2 Ib. The color indices are very similar, and if we use the relation between main sequence and giants, and allow for the difference in absolute magnitude between -3.1 and -5.5, we find an expected spectral type of M1.5 Ib for the globular cluster stars. Here we have a quantitative discrepancy, G8 observed and M1.5 expected.

The situation is similar in M 13, and in M 92 it is even more extreme. In M 13 we should expect a spectral type of K3, and on the basis of the strength of H$_\gamma$, we get G2 Ib.

As for the stars on the horizontal branch, their $U - B$, $B - V$ curves follow the curve for normal stars. Their spectra were usually classified as normal B8 and A0 stars, but the situation changed when they were studied with larger dispersion. According to Guido Münch, who has done most of this work, the spectral types based on the hydrogen lines are usually much later than would be expected from the color indices. A star which one would expect to be of type O from the color index turns out to have a spectrum something like B2. For example, the faintest star that Münch has studied in M 13 has $M_V + 3.0$ and color index $-0^m.40$, and here one finds very strong hydrogen lines, quite inconsistent with the very early color index. Where one would expect weak hydrogen lines, he finds very strong ones.

A very similar case is the star BD + 25°2534, of the ninth magnitude, in the galactic halo. It was first found to be a very blue star in the 1920's by Malmquist, when he studied color indices and spectra in a region around the north galactic pole. The spectroscopists reported that it was a normal B8 star, which was remarkable, because the star was one of the bluest stars that Malmquist found. During the last few years the star was rediscovered, first in the search for blue stars with the 48-inch Schmidt at Palomar, and at the same time by Stock and Slettebak on objective-prism plates taken with a Schmidt; it was shown to be one of the bluest stars in the sky.

When Münch took spectra with larger dispersion, he again found strong lines of hydrogen, surprising in view of the very blue color index. But his plates also showed He II 4686 and some weaker lines of He I; on the basis of the He II line the star would be called an O star. Münch favors the interpretation that we are dealing with a star that has a very steep temperature gradient, or else that the star's atmosphere is surrounded by a shell, in which the Balmer lines and a strong Balmer jump are produced.

The spectrum scanner devised by Code can cover the whole spectrum from 3300 A to about 10,000 A — a beautiful method for spectrophotometry because the photometric scale is well determined. Code found that BD + 25°2534 has a spectral distribution consistent with that of an O star in the red and infrared. In the blue and ultraviolet the temperature is apparently less, but from 5700 A to 10,000 A the whole distribution of energy can be represented by a single gradient. This result is in agreement with the suggestion of Münch that the star itself has a very high temperature, but that the spectral type is faked by the strong hydrogen lines in the outer atmosphere or in the shell.

Münch is investigating a number of extremely blue stars, of absolute magnitude between +3 and +6, that have been found in the galactic halo. One of them, which is typical of a group, is BD + 25°4655, remarkable because the helium lines are exceedingly strong, and the star must be very rich in helium relative to hydrogen. In addition, many lines of N II, N III and Ne II are remarkably conspicuous. On the other hand, only on the very best plates can one occasionally see bare traces of the lines of oxygen and carbon, although many of these lines in several stages of ionization lie in the same range of wavelength. The results of a recent quantitative analysis of two of these stars are very remarkable.

The ratio H:He for these stars is found to be 1:4, whereas for normal stars it is more nearly 4:1. The results for the heavy elements are most remarkable (Table 8), and

Table 8. Abundance of heavy elements
in two blue stars.

Atom	Abundance
N	100
C	0.5
O	6
Ne	20
Si	6

support the idea that the stars are very old; helium is now the important element. The electron temperature is about 35,000°K, the surface gravity of the order of $10^{6.8}$, and the diameters of the order of 0.1 solar radius. The masses are of course unknown, but absolute magnitudes from about +3 to +6 are compatible with masses between 15 solar masses and 1 solar mass. Here we have one of the oldest stars on the downward turn of the horizontal branch, and Münch has very similar data for six other stars.

Code has studied two red giants in M 13, and finds deviations in the opposite direction to those found for the blue star. Again the gradient is compatible with a single temperature, but that temperature is very much lower than that to be expected from spectral type and color index. Code suggests that we have here a blanketing effect, which is practically zero in the infrared region, small in the blue region, larger in the ultraviolet. This would mean that the stars have moved off the normal $U-B, B-V$ curve because they have a large $U-V$ excess and a smaller $B-V$ excess. But whether blanketing alone is responsible remains to be seen.

I now turn to the integrated spectra of globular clusters, which show that the apparently homogeneous group of globular clusters, with minor differences of chemical composition, can be broken up again into definite groups; that is what always happens when a group is studied closely.

Two years ago, W. W. Morgan began to derive the spectral types from the integrated spectra of globular clusters at the McDonald Observatory. If we know the luminosity function of a cluster, we can say which stars contribute to the integrated spectrum. For instance, the investigation of the luminosity function of M 3 by Sandage has shown that practically all the visible light comes from the three brightest magnitudes. The horizontal branch may contribute a little — in some cases an appreciable amount — but practically 97 percent of the light of such a cluster comes from the first three magnitudes. Consequently, anything that we observe in the integrated spectrum is more or less a common property of the giants.

For the integrated spectra Morgan again found that the spectral type depends on the criterion used. Table 9 summarizes the results that he obtained from the ratio of the G band to $H\gamma$ ($CH/H\gamma$); the intensity of $H\gamma$ alone; and the metallic lines (Fe I).

If we compare the spectral types derived from $H\gamma$ with those derived from the metallic lines for the first eight clusters, we see that there is in general a difference of 0.7 to 0.8 of a spectral class, in the sense that the spectral type derived from metallic lines are always earlier.

However, for the three last clusters, all three criteria give

Table 9.   Integrated spectra of globular clusters (Morgan).

Cluster	CH/Hγ	Hγ	Fe I
M 92	F2	F6	—
M 15	F3	F6	—
M 53	F4	F6	<F0
M 5	F5	F8	F0
M 13	F5	F8	F0
M 3	F7	F8	F1
NGC 6229	F7	F8	F0
ω Cen	F7	F8	F0
NGC 6356	G5	G2	G5
NGC 6637	G7	G8	G2
NGC 6440	G5	G5	G2

essentially consistent results. Now note a remarkable thing: the mean distance from the galactic plane, $|\bar{z}|$, is 8.3 kpc for the first eight clusters, and only 1.0 kpc for the last three. The reason that the second group is so small is that most clusters of this kind are in low galactic latitude, and either are involved in dense clouds or are in very rich regions along the galactic equator.

Morgan concluded that the last three clusters belong to a second group of clusters, differing in chemical composition from the others, and that they contain normal stars, for the same criteria would lead to the same results for normal stars. He expressed the opinion that these globular clusters of normal composition are concentrated to the galactic center.

A similar suggestion had been made in 1948 by Wilhelm Becker, on a different basis. When Mayall published the radial velocities of globular clusters, he classified their spectra; he used a mixed criterion of H lines and metallic lines, and found the large range from A5 to G5 for the integrated spectra. Becker called attention to the fact that the clusters which Mayall classified as G had a different distribution over the sky from those with earlier spectra, and were restricted to a small sector of longitude between 315° and 5° (including the galactic center at 327°), and to low galactic latitude.

Table 10 summarizes some of the data bearing on these

Table 10.   Integrated spectra of globular clusters (Mayall).

Spectrum	n	$|\bar{b}|$	$|\bar{z}|$ (kpc)
A5–A9	6	39°.0 ± 9°.8	
F0–F4	8	31 .5 ± 8 .1	
F5–F9	12	27 .3 ± 5 .6	
G0–G2	13	17 .0 ± 3 .1	
G3–G5	10	8 .5 ± 1 .5	
A5–G0	31	29 .3	8.1
G1–G2	8	15 .6	4.4
G3–G5	10	8 .5	1.1

G-type globular clusters. The spectra determined by Mayall are divided into groups, and the mean galactic latitudes, $|\bar{b}|$, and distances from the galactic plane, $|\bar{z}|$ are given. The first groups show very little concentration to the plane of the galaxy, but the G3–G5 group shows considerable concentration. It turns out that you can include all spectra up to G0 in the first group. In the second part of the table, the G3–G5 group really sticks out. The intermediate group (G1–G2) may possibly be a truly intermediate group, or it may result from the fact that the spectrum is not quite enough to assign the cluster to its proper group; with these data it cannot be decided. In determining the values of $|\bar{z}$

I tried to make allowance for the absorption, using Lohmann's values, which, though not accurate, should be sufficient for this purpose.

When we subdivide in three zones of latitude, the thing becomes very clear. The A5–G0 clusters belong to the galactic halo; the northern cap contains 9, the next zone 13, which of course means that they are concentrated to the galactic center. The cause of the drop in numbers in the 0°–19° zone is, of course, the absorption in the galactic plane. You note that there are none of the intermediate group (G1–G2) in the northern cap.

Table 11 shows in a very striking way how one group is

Table 11.   Distances of globular clusters from galactic plane.

$|b|$	A5–G0		G1–G2		G3–G5	
	n	$|z|$	n	$|z|$	n	$|z|$
0°–19°	11	2.1	5	2.4	10	1.1
20 –42	13	8.9	3	7.2	0	
43 –90	6	13.5	0		0	

distinguished by its restriction to small $|z|$. But it is not true that this group is concentrated at the galactic center. Several clusters lying outside the central region are now known. One of them is NGC 6838, M 71, which was added to the list of globulars by Mayall, being included in both Shapley's and Trumpler's lists of galactic clusters. Fortunately the large radial velocity determined by Mayall at once showed that it is a globular cluster. Mrs. Hogg has carefully searched this cluster for variable stars and has found none. Another is NGC 2158, which obviously belongs to this group, although it is actually in the anticenter region. So it is not true that these clusters are confined to the central region of the Galaxy; they show a strong concentration toward the central region, but they are obviously distributed through the disk. So there is good reason to call them a disk population.

Even the globular clusters have thus been divided into two groups: one an essentially spherical system, strongly concentrated toward the center; the other a disk system, again with strong concentration toward the center within the disk. They differ in chemical composition, in the sense that those in the halo have low metal content, but with variations (for example, bewteen M 92 and M 3), whereas those in the disk have practically normal composition, no differences being known at present. We are of course now greatly interested in the make-up of these disk clusters. Since the only thing of a similar sort that we know is M 67, it is tempting to think that we have a similar case here, or at least a closely related one.

These globular clusters have integrated spectral types about G2, G3, or G5. They must in any case be old clusters; if they were relatively young like the Hyades or Praesepe they would contain very few giants, and might easily have very early spectral types, such as A5. This must mean that the giants pass in some way through the subgiant into the giant region. They do not necessarily resemble M 67, they may be intermediate cases; that must be decided from the spectra. But they must be systems in which the giants come in from below: cases like NGC 752 or M 67.

The crucial test, the observation of the color-magnitude diagrams, will be quite difficult, because the number of clusters that promise to be manageable is very small. Probably the most promising will be M 71. Although NGC

6356 seemed to be the most favorable, the absorption is so heavy that we cannot reach the main sequence. Perhaps NGC 2518 will also prove to be possible.

Let us survey the clusters of the galaxy. First we have the halo system of the globular clusters; we have good reason to believe that these consist of old stars with low metal content. When we turn to the disk clusters we find a group that have up to the present been called globular clusters, but that have normal metal content. Finally, among the younger clusters up to M 67, we have a whole gamut of ages, but with very small differences of chemical composition.

Consider the plane of the galaxy. The open clusters are very strongly concentrated to it. If they are subdivided according to age groups, the youngest will show a terrifically strong concentration. In adition, within this rather thin layer, we have this group of G3–G5 clusters which have kinematic characteristics rather like those of the high-velocity stars, and are distributed throughout the disk. The value of 1.5 kpc that I gave for the mean $|z|$ component is too large, because in obtaining it I assumed the same upper limit of luminosity, -3^M, as for the globular clusters, whereas if they are normal systems related to M 67 the upper limit will be not -3^M but 0^M, so the value 1.5 kpc should be divided by a factor of 4, giving for $|z|$ something of the order of 0.4 kpc.

When we discuss the Galaxy, we shall see that probably we shall have to understand this distribution as a consequence of the shrinking of the gas layer that has produced stars, and has now become very thin. How does the distribution of clusters represent the history of this shrinking layer of gas? I think we shall see later that it makes sense to order the objects in the Galaxy according to this shrinking-gas picture.

The G3–G5 clusters are strongly concentrated to the plane of the Galaxy. In this way they closely resemble the high-velocity stars, and share a characteristic of these stars that is usually completely forgotten — their strong concentration both toward the plane of the Galaxy and toward the galactic center.

These globular and open clusters give us for the first time the picture that the disk of our Galaxy is made up of objects of many kinds, differing in age as well as in chemical composition. When we turn to a discussion of the Galaxy, we shall have to see whether we can make sense of this picture, whether we can, at least to a first approximation, arrive at an idea of its history. Of course we are at the very beginning of this discussion, but it is very tempting to look at things from this point of view.

CHAPTER 14

VARIABLE STARS

IN GLOBULAR CLUSTERS

When we discussed galactic clusters we encountered only three types of physical variables. In h and χ Persei there are irregular red variables, supergiants of spectral type Mo to M3. Then there are Type I (classical) Cepheids in a number of clusters, the richest being probably NGC 1866 in the Large Magellanic Cloud. And finally there are the T Tauri variables in very young clusters, and the associated flare stars, for instance, at the very faint end of the Pleiades. In contrast, the globular clusters contain a large variety of physical variables.

In order to assemble the variable stars that are members of Population II, I went carefully through the published lists of variable stars in globular clusters, especially Mrs. Hogg's list, and finally got a simon-pure list. I shall mention the cases that are still in doubt.

What is the shortest period for real cluster-type variables in globular clusters? The shortest periods actually observed

Table 12. Periods of cluster-type variable
stars in globular clusters.

Cluster	Star no.	Period (d)
NGC 6522	6	0.19
	5	.22
	3	.22
M 9	10	.24
M 4	37	.24
	41	.25

are given in Table 12, and it turns out that $0^d.2$ is about the real lower limit. Stars with periods over $0^d.25$ set in large numbers. The cluster NGC 6522 lies in the neighborhood of the galactic center.

There is a well-known group of short-short-period variables, such as SX Phe, AI Vel, and CY Aqr, with periods starting at $0^d.05$, running up to $0^d.19$ for δ Sct. These stars obviously belong to a very different group. Our first information about their absolute magnitude shows that they lie in the interval from about $+4^M$ to $+2^M.2$. They seem to show a period-spectrum relation in the sense that those of shortest period are of about type A2, those of longest period, A9 to F0 or F1. Those of shortest period have the lowest luminosities.

Two such stars have been reported in globular clusters. The one in M 56 can be ruled out right away; it is three times as distant from the center of the cluster as any of the other cluster-type variables, and the cluster is involved in a rich field. The star in ω Centauri with a period of $0^d.06$ can also be ruled out, because it would have an absolute magnitude of $-0^M.4$ if it were a member of the cluster, whereas we expect something nearer to $+4^M$. Variables of the SX Phe type are clearly so frequent in space that it would not be surprising to find one in an angular extent as large as that of ω Centauri. The case for short-short-period variables in globular clusters rests on this one star, and although it lies within the main body of the cluster, I rule it out on grounds of luminosity.

The present data on variable stars in globular clusters are summarized in Table 13. The number of known RR

Table 13.   Variable stars in globular clusters.

Type	No. of stars
RR Lyrae	>1000
Type II Cepheids	30
RV Tauri Stars	4
Semiregular and irregular	32
100^d variables, small amplitude	4
Mira stars, period $\approx$ 200^d	5
Novae	1
SS Cygni stars	?
Eclipsing stars	3

Lyrae stars in larger than 1000 now, and the figure would certainly be doubled if all the material was worked up. There are 874 of known period, and the magnitude ranges of a large number of others show that they are cluster-type variables.

I have used these data, which I consider reliable, to form a period-frequency distribution for the stars of period greater than a day (Fig. 23). The second maximum, which is very conspicuous, contains the W Virginis stars, in which I have included both the peaked and flat-topped light curves. For these stars it was simple to decide whether they

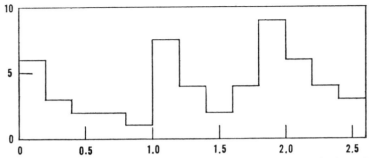

FIG. 23. Frequency of logarithm of period (abscissa) for intrinsic variables in globular clusters with periods over 1 day.

were members of the clusters; many of the radial velocities have been checked, mainly by Joy, and a few by Wallerstein. For the W Virginis group the star of shortest period is No. 7 in M 14 ($13^d.6$); the longest is No. 6 in M 2 ($19^d.3$).

It is quite clear that the group with a maximum near $2^d.5$ should be counted with the cluster-type variables, where Joy always wanted to put them; if we were to include the cluster-type variables with periods less than a day, this maximum would of course have a tremendously high peak.

There are four RV Tauri stars in the list: No. 1, ω Cen ($58^d.7$); No. 4, M 4 (98^d); No. 6, M 56 (90^d); No. 11, M 2 ($67^d.1$). These periods should, of course, be halved, because as we shall see the basic spectroscopic period is $P/2$, but here I give the doubled period that is usually given. There are probably some RV Tauri stars hidden among the semiregular and irregular variables; the only way to sort them out at present would be by their spectra, as Joy has done for other stars. Except for the star in ω Centauri, the stars I have listed have been checked for light curve and spectrum, and they are bona fide RV Tauri stars.

The group of semiregular and irregular variables contains 32 stars, about as many as the Type II Cepheids. The spectra of a number of them, investigated by Joy, are between G0 and K2. Their light curves are irregular, but we may suspect that some of them are RV Tauri stars because Joy observed emission lines for about 5 percent of them, and we know that RV Tauri stars, like W Virginis stars, show bright hydrogen lines on the ascending branch of the light curve. The percentage must be small, and one way of sorting out these stars would be to observe them spectroscopically with larger dispersion than Joy used.

The amplitudes of these irregular and semiregular variables are of the order of a magnitude, and their median photographic absolute magnitude is about $-1^M.40$, which may be provisionally adopted as characteristic for these stars. They usually lie near the tip of the giant branch, and their spectra are usually variable, with extremes from G5 to M0 or M1 with no emission lines.

Figure 19 shows the color-magnitude diagrams of four globular clusters that contain irregular or semiregular variables. No. 95 in M 3 has a period of 103 or 105 days, a color index of $+1^m.6$, an amplitude of about a magnitude, and a very regular light curve. In M 13 there are two semiregular variables, with a mean absolute magnitude, integrated over the cycle, of $-1^M.4$. In M 5 the variable lies in the extension of the giant branch, and in M 10 it lies half a magnitude higher. These stars always lie in the neighborhood of the giant branch.

The long-period variables observed in globular clusters fall into two groups. One group contains stars of small amplitude and periods around 100^d, with regular and symmetrical light curves. Star No. 95 in M 3 has a period of 105^d, amplitude $1^m.0$, median photographic magnitude $-1^M.7$. Joy has derived the radial velocity curve, and the phase relations are about the same as for the Type II

Cepheids and RV Tauri stars. Only four stars of this group are known, but a number may be hidden among the semi-regular and irregular variables. The group is very important; we shall see later that its members are quite frequent in the galactic nucleus. The globular clusters do not give us much information on the range of periods, but in the galactic nucleus the range is from about 70^d to 150^d, with the most frequent period around 100^d, even in the galactic nucleus.

The second group of long-period variables, the Mira stars, are a much-disputed subject. Five are known at the present time, of which two (No. 2 and No. 42) are in ω Centauri. The period of No. 42 is $149^d.4$, and its absolute photographic magnitude at maximum is $-2^M.1$; there are no uncertainties about this star. The period of No. 2 is usually given as 484^d, but Martin, in his extensive work on ω Centauri, pointed out that the Leiden observations indicated a period of 242^d. Bailey's early data, on which the doubled period was based, show that at the time of his observations the star had alternately high and low maxima, which we know today is an occasional feature of a number of Mira variables. There is no doubt that 242^d is the true period here; the highest maxima reach $12^m.2$, the lowest, $13^m.6$, and the average magnitude at maximum is $-1^M.7$ photographic. The minimum brightness is 16.0, so the amplitude varies between $2^m.3$ and 4^m.

To obtain the absolute visual magnitudes at maximum for the long-period variables in ω Centauri we add the color index to the photographic magnitudes, and obtain $-3^M.5$ if we use the cluster-type variables as zero point. The absolute visual magnitudes for Mira stars (based on proper-motion data, which are not very reliable) are somewhat lower, lying around -2^M or $-2^M.5$.

In 47 Tucanae there are three Mira variables: No. 1 ($212^d.4$), No. 2 ($202^d.8$), and No. 3 ($192^d.3$). The observations of their spectra, made at Pretoria, show that they are Me stars with all the characteristics of long-period variables. Their magnitudes are all within a few tenths of $11^m.47$ photographic. The cluster 47 Tucanae is one of the richest known; if it contains cluster-type variables their number is exceedingly small.

In default of cluster-type variables to set the zero point, we can use the brightest cluster stars, which in normal globular clusters are $1^m.5$ brighter than the cluster-type variables. But the remarkable thing is that the average absolute magnitude of the Mira stars in 47 Tucanae, on this basis, is $-3^M.2$ photographic as against $-1^M.9$ photographic in ω Centauri; if again we add the color index of $1^m.3$ we get $-4^M.5$ visual. This does not make sense, because if Mira stars are so bright we should find them galore on 200-inch or 100-inch plates of the Andromeda Nebula, whereas we can distinguish them in the blue only under the very best conditions.

It is obvious how we can bring things into agreement. We should have to apply a correction of about $1^m.3$ to the magnitudes derived on the assumption that 47 Tucanae is a standard cluster. We should have to assume that the giant branch in 47 Tucanae reaches not up to $-1^M.5$, but only up to $-0^M.2$. Here we come close to our friend M 67, which is just as old, but contains more metals, and therefore reaches only to 0^M.

Thackeray reported on two studies of 47 Tucanae at the Rome Conference. The first, the color-magnitude diagram, has not yet been carried very far, but we can say definitely that the giant branch extends considerably farther to the red than that of M 3; it is shifted in the direction of the galactic halo clusters. There is no reason to assume that this is an effect of obscuration, since 47 Tucanae is projected against the edge of the Small Magellanic Cloud, and Elsasser's recent photoelectric work on the Clouds shows that the color in this part of them is very uniform.

The second study is a spectrographic investigation by Feast of the giants in 47 Tucanae. In one way they are very different from stars in normal globular clusters, in which, although the colors indicate that they should reach Class M, the spectra are actually always K0, K2. But here for the first time we find more than half a dozen M stars, up to M 3, a remarkable thing. The whole thing is shifted over, and now apparently the spectra agree with the colors, which means that the stars must be close to normal. With a dispersion of 20 A/mm, Feast tried to determine the luminosities of these M stars and of one very blue star which lies above the horizontal branch, and found that these luminosity criteria required a correction of $+1^m.2$. These luminosity criteria have worked very well even for extreme halo stars, although of course the accuracy is not very high, giving only the luminosity classes I, II, III. I am sure that when we have the color-magnitude diagram we shall find that in 47 Tucanae we are not dealing with a halo cluster.

Another indication in the same direction is that the integrated color index for 47 Tucanae, measured by both Irwin and Oosterhoff, is $+0^m.98$, whereas the normal integrated color for the globular clusters of the galactic halo is $+0^m.68$. Elsasser's very careful photoelectric photometry shows that there is no absorption surrounding 47 Tucanae, unless you place it right in the center of the cluster, which would be perfectly crazy. Here we have another indication that 47 Tucanae is not a normal globular cluster, and in agreement with this we find that, if we assume that the upper limit of brightness of the giant branch is only $-0^M.2$, the z-component for 47 Tucanae is of the order of 2.7 kpc.

If we regard the absolute magnitudes in 47 Tucanae as really fixed, we now have absolute magnitudes for three more Mira stars in globular clusters, bringing the number to five in all. The range of periods is from 149^d to 242^d, and the mean period just about 200^d. You recall that the 200-day group of Mira variables in our own Galaxy has a very high space velocity. We shall see later also that this group is very highly concentrated to the region of the galactic center. So at the present time our best value for the absolute magnitude at maximum is about $-1^M.9$ photographic for this group, and this value makes sense with the results of our attempts to find long-period variables in other galaxies, especially M 31.

It is very difficult to get data on the luminosities of long-period variables from proper motions and radial velocities. Therefore it would be very valuable to investigate other long-period variables in globular clusters. One of these is No. 68 in NGC 3201; it has a large amplitude, is a member of the cluster, and should be a long-period variable. Other cases are No. 1 in NGC 6171 and No. 7 in NGC 6712; both clusters contain cluster-type variables. For the former the maximum photographic magnitude must be about $-2^M.0$, and for the latter, about $-2^M.3$. It would be very important to get the periods for these two stars. Finally, No. 1 in NGC 6541 has a very large amplitude. This would bring up to nine the absolute magnitudes of Mira stars in globular clusters.

I have mentioned the RV Tauri stars in globular clusters, but we also have RV Tauri stars in our Galaxy, and Joy has made a very careful study of them. He found that he could arrange them in two groups by their spectra, and when he studied the radial velocities it turned out that there was a high-velocity group, and a low-velocity group (about 40 km/sec) belonging to Population I. There may be differences in period, and other small differences in behavior, between the two groups.

We now come to the novae. The nova T Scorpii, which appeared in 1860 in the globular cluster M 80, is exceedingly well observed, and there is not the slightest doubt that it belongs to the cluster. The coordinates of the nova, referred to the center of the cluster, are $+4''$ and $-3''$. The relations between light curve and absolute magnitude at maximum, found by Arp in his study of novae, lead to $M_v = -8.5$, and a distance modulus of $15^m.7$ for M 80, whereas the distance modulus derived from the cluster-type variables is $15^m.93$. About Nova Sagittarii 1943, in NGC 6553, I am doubtful. It also reached a high absolute magnitude, $-8^M.35$, but it is on the outskirts of the cluster, and its membership is uncertain.

Two possible SS Cygni stars have been found in globular clusters: No. 101, M5 (maximum $17^m.15$, $+2^M.1$, mean amplitude $4^m.3$) and No. 4, M 30 ($+0^M.9$). The latter is excessively bright for an SS Cygni star and is far from the center of the cluster.

Three eclipsing stars have been reliably assigned to globular clusters, ranging in period from $1^d.1$ to $1^d.7$, in amplitude from $0^m.7$ to $1^m.8$, and in luminosity from $-0^M.5$ to $-1^M.1$.

In the last chapter we saw that the globular clusters must be subdivided into two groups: the halo clusters, which are apparently poor in metals, and the clusters of the disk, which seem to have normal metal content. Do the globular clusters of the halo and the disk differ in respect to their variable stars? Unfortunately we cannot yet say anything definite, because of the ten clusters that Mayall classified as G3 to G5 seven have never been searched for variable stars. They are far south, and must be investigated from the Southern Hemisphere, and because of their low galactic latitudes they are either projected against rich star clouds or (more commonly) embedded in nebulosities. For example NGC 6440, one of the disk clusters, is so heavily obscured that we can reach only the brightest stars photographically, and cannot hope to reach the variables.

However, three of Mayall's disk clusters have been thoroughly searched. In NGC 6838, M 71, which is quite easily accessible in Sagitta, no variables whatsoever have been found in a recent thorough search by Mrs. Hogg. The two others give contradictory results. Mayall classified NGC 6723 as G3, so it should belong to the group, but it contains 19 cluster-type variables and its z-component is -3.5 kpc. In NGC 6712 (G4) there seem to be 6 cluster-type stars; though no light curves have yet been derived, their magnitudes and ranges suggest that they are of this type; the z-component of the cluster is -0.8 kpc.

The large distance of NGC 6723 from the galactic plane makes it a little bit suspect, and it is quite possible that it may be G2 or G1 and not to be counted as a disk cluster. But the variables in NGC 6712 should be investigated to see whether they are really cluster-type stars. The cluster is also interesting because it obviously contains a long-period variable of undetermined period. Its distance modulus is $16^m.5$.

I have suggested that these globular clusters of the disk might be of M 67 type. Cluster-type variables would hardly be expected in M 67, since the brightest stars at the tip of its giant branch have only magnitude 0^M, and the stars that have moved off to the blue side are a magnitude fainter. To make it certain, I asked Rosino to carry out a thorough search of M 67 for variable stars. He made it several years ago, and found not a single physical variable. It is, of course, quite possible that the disk clusters are intermediate between the normal globular clusters and the M 67 type. It will be very interesting to see how far composition is related to the appearance of cluster-type variables.

I now turn to the so-called Cepheids in globular clusters. When Bailey made his historical investigation of the globular clusters, he found, besides the numerous cluster-type variables, a very small number of Cepheids with periods greater than a day. Since the light curves looked somewhat similar, at least for periods shorter than 10 days, it was generally assumed that these stars were ordinary Cepheids. Later, however, it turned out that there were definite differences. One of the first things that became obvious was that the period-frequency distribution of these "Type II Cepheids" was very different from that of the classical Cepheids, with a minimum between 5 and 10 days.

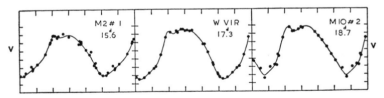

FIG. 24. Comparison of V light curve of W Virginis with those of two Cepheids in globular clusters.

The type II Cepheids with periods between 10 and 20 days had different light curves from the classical Cepheids in the same period interval (Fig. 24). The variables with periods between 10 and 20 days in globular clusters resembled W Virginis in our own Galaxy, and the whole group has been called the W Virginis group.

Finally it turned out that the globular clusters, in addition to Cepheids with periods between 1 and 10 days (whose light curves cannot be easily distinguished from those of classical Cepheids) and the W Virginis stars, contain also RV Tauri stars. Usually the RV Tauri stars are very much brighter than the other cluster stars, and they often lie well outside the clusters. In order to get better data on these stars, Arp reinvestigated the whole series of variable stars with periods greater than 1 day in a number of globular clusters. He always connected their magnitudes with those of cluster-type variables in the same cluster, and so his results are referred to the zero point of the cluster-type variables.

It turns out that there are two groups of W Virginis stars; in one group there is a long halt in the light curve at maximum, and I think the "flat-topped W Virginis stars" is a good name for them; the phase of the mean of the flat top is about 0.18, and it lies on the average about $0^m.15$ below the maximum. The second group have a halt on the descending branch, at a phase of about 0.39, which lies about $0^m.75$ below the maximum. The type with the halt on the descending branch seems to be very regular both in period and light curve, but changes of period and light curve are quite frequent among the flat tops.

Both types of W Virginis star always show bright hydrogen lines as the star increases in brightness to maximum. This is also a very characteristic feature of RV Tauri stars, but is not observed in classical Cepheids. In the latter we occasionally observe bright H and K lines, but that is very different from these bright hydrogen lines. Finally it was found, when the Type II Cepheids were studied with larger dispersion, that all of them showed a doubling of the absorption lines near maximum.

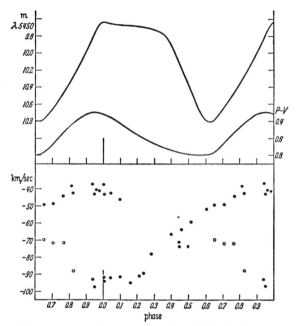

FIG. 25. Light curve, color curve, and radial velocity curves of W Virginis from absorption lines (*solid circles*) and radial velocity of emission lines (*open circles*).

Recent investigations of Type I (classical) and Type II Cepheids have shown that the kinematic mechanism is different in the two cases. If we consider the velocity curve of a Type I Cepheid and assume that the star is pulsating, it is very easy to locate the critical phases. The velocity curve crosses the mean line at maximum and minimum size, because then the velocity of the surface is the same as the velocity of the system. So we obtain the variation of the radius as a function of time by an integration of the velocity curve.

As an example of a Type II Cepheid we can take W Virginis, in our own Galaxy, which has been very well investigated. When the star is at maximum, we observe two sets of absorption lines, one displaced to the red, the other to the violet (Fig. 25). In any one cycle, a set of violet-displaced lines appears. This set of lines moves toward the red until the next maximum, and then its lines grow quite faint and disappear. At each maximum of a Type II Cepheid a layer of the star is suddenly endowed with very high velocity, expands to maximum size, and then falls back. If you compare with the Type I Cepheid you will note the great difference; the corresponding part of the velocity curve extends only from maximum to about minimum, whereas for the Type II star it covers the whole cycle, from maximum to maximum. For the classical Cepheids the mechanism is such that the surface completes the whole cycle from maximum diameter to minimum diameter, but for the Type II Cepheids the layer is ejected, and does not complete the cycle. If now you integrate the velocity curve, you find that the star's volume increases to a maximum, and then falls back, but before it reaches a minimum the material hits the next layer that has been ejected; it almost looks as if two things collide. I think in the case of the classical Cepheid there is synchronism between action of this sort and pulsation of the star, whereas in the case of the Type II Cepheid there would be no synchronism. So there is a fundamental difference between the pulsation mechanisms in the two cases.

The γ-axis can be fixed by Joy's observations of four W Virginis stars and three RV Tauri stars in globular clusters. The mean velocities are identical with the velocities of the clusters in which the stars are situated. The dispersion of the velocities is very small, and so we are justified in assuming this axis as representing the star's velocity.

For the RV Tauri stars I will use the data from Abt's

paper on U Mon. The periods of RV Tauri stars are quite variable, and sometimes high and low minima alternate. The velocity curves show something similar to what we saw for W Vir: every time at maximum light you again see two sets of absorption lines. There is one qualification; very often the period given for such a star is the interval between similar minima, because the minima often alternate in depth, but the radial-velocity curves show clearly that the true period should be one-half of this. Here again, as for the W Virginis stars, we do not have synchronism between the mechanism that produces the oscillation and the oscillation itself. The period of free oscillation of the star is obviously longer than the period of this mechanism; one finds that the ratio of the period of free oscillation to the period of forced vibration is of the order of 1.5.

We now consider the period-luminosity relation for the Type II Cepheids. If all the differences between Type I and Type II Cepheids had been known at the beginning, we should not have attempted to squeeze two very different things into the same bed of Procrustes, for that is what happened. In addition to the differences that have been mentioned, we should remember that while, to the best of our knowledge, the Type I Cepheids conform to the mass-luminosity law, we now strictly have to assume that all Type II Cepheids have the same mass, 1.2 solar masses. I think that it is out of the question to unify the two groups of Cepheids. True, they are both oscillating; but the type of oscillation is different and the masses are different too.

If one attempts to form a period-luminosity relation for the Type II Cepheids in globular clusters by plotting their mean magnitudes referred to the luminosity of the cluster-type variables, one gets rather a confused picture. We have every reason to believe that there must be a close connection between the W Virginis and RV Tauri stars, and I tried to represent them by two parallel lines, but the deviations are large. After trying many other ways of representing the data, I finally plotted absolute magnitudes (both photographic and photovisual) at *minimum* against period. In both cases one gets a remarkable result. Now the W Virginis and RV Tauri stars form a unified group. I have included some more recent observations of RV Tauri stars with still longer periods. The whole scatter in the magnitude-luminosity diagram turns out to have come from the amplitudes. But what justification do we have for considering that minimum brightness should give a better correlation than average or maximum brightness?

CHAPTER 15

ELLIPTICAL GALAXIES

AFTER our detailed discussion of the color-magnitude diagrams of star clusters in the two populations, we turn to the galaxies, and apply our evolutionary views to their interpretation.

In the detailed study of elliptical galaxies we are restricted to the relatively small number within our own Local Group; the brightest of these are the companions of the Andromeda Nebula, and there are about half a dozen dwarf galaxies of the Sculptor type. Four of the Sculptor-type galaxies are so close that we can quite easily reach the cluster-type variables with modern equipment, and can also hope to extend the color-magnitude diagram to the horizontal branch, or even fainter, and in recent years I have kept all four under study. The Draco system, the Ursa Minor system, and the Sculptor system are the closest; the latter is being studied at Pretoria. The Leo system is in very high galactic latitude, and about twice as distant as the others, but easy to observe.

The investigation of the Draco system is now completed. The observations were made with the 200-inch, which meant that on one plate we could get only the central part of the system. Observations with the 100-inch would have been sufficient, but its mounting is such that it cannot reach Declination $+67°$. The apparent diameter of the Draco system is 48′, and the 5 × 7-inch plates at the 200-inch cover something like 8′ × 10′. The whole area searched was square, about 20′ × 20′; a rather thorough survey was made for variable stars by the intercomparison of about 30 pairs of plates. All the 134 variables found are cluster-type variables, if we count as cluster-type all the variables with periods up to 2½ days, for there are two or three with periods greater than 1 day.

A number of fields surrounding the central field were observed occasionally with the 200-inch in order to get an idea of the extent of the system from the cluster-type variables themselves; 8 or 10 plates for each of these fields were searched for variables. These side areas furnished 53 variables; they were of course not completely searched, but they give an idea of how far the cluster-type variables extend. I have drawn a circle around the variables with a diameter of 48′; I think it is quite conservative — one of the variables is even outside the circle. On the Schmidt plates one can see that the Draco system is very difficult to disentangle from the foreground, by star counts for instance, because of the relatively low galactic latitude; the same is true of the Ursa Minor system. It would not have been worth while to make star counts.

Miss Swope has found 134 variables in the central field, and has derived their periods and light curves. The light curves are quite normal; they start from $0^d.55$ (actually not quite the shortest period) and run up to $1^d.59$. In one way the Draco system is very remarkable; true type c curves do not occur in it. We always find two types of light curves in globular clusters: type c with the form of a sine curve, and the asymmetric curve of large amplitude. But the type c curves are missing from the Draco system; they do not even occur for the shortest periods. Instead we find very regular light curves for an interval of period, and then suddenly the light curves show a large scatter; then again there is an interval of period where the light curves are regular, then again they are scattered, and so on. The amplitudes may be small, but the light curves are always asymmetric. The light curves change in a recurrent way that we do not understand, and in this respect the variables in the Draco system are quite different from those in globular clusters.

The frequency distribution of the periods in the Draco system is very remarkable. The periods start at $0^d.3$, but the light curves are already asymmetric. And 75 to 80 percent of the periods lie in a very narrow interval ($0^d.55$ to $0^d.70$), after which they fall off, so there is a very pronounced peak in period frequency. This again is something unusual, which has not been found in globular clusters. The mean period for all the variables in the Draco system is $0^d.614$; for the globular clusters Oosterhoff found two groups, one with a frequency maximum at $0^d.525$, the other at $0^d.625$.

The color-magnitude diagram for the Draco system has all the features shown by globular clusters; there are plenty of foreground stars, for the galactic latitude is 28°. There is a giant branch running into a subgiant branch, a horizonal branch, and a cluster-type gap. On the $B-V$ system this gap lies a little below 20^m, and the giant branch extends to about 17^m, so the brightest stars are −3^M photovisually if we adopt o^M for the cluster-type variables. In its general features the diagram is the same as those of globular clusters.

But the Draco system also shows some remarkable differences as compared to the globular clusters, for example the great strength of the horizontal branch toward the red side of the cluster-type gap, and the extreme weakness at the blue side of the gap. The difference can be seen by comparing with the diagrams for M 3 and M 5 (Fig. 19), where there is relative weakness on the red side of the gap, great richness on the blue side. So there are real differences, although the general features are the same.

Most of the stars that are scattered over the diagram will be field stars, but there is a star at 17^m.2, color index −0^m.25, just about as high as the giant branch, that is probably a member of the system. This is one of the bright blue stars that we find again and again in globular clusters, and especially in Sculptor-type systems. Its absolute magnitude is −3, and it is of special interest.

In one way this diagram of an elliptical galaxy gives a truer representation than those of globular clusters. We cannot observe the central area in M 3 and M 5, which are very condensed clusters, but in the Draco system every star has been measured, and we have a real representation of the central area. Unfortunately we have no detailed studies of poor globular clusters; all the work has been done on the rich ones. The ideal comparison would be with a poor globular cluster such as NGC 5053, where we can observe right into the center. It is possible that there are differences in the relative strengths of the giant branch and the horizontal branch; by simply comparing the diagrams of M 3 and M 5 with that for the Draco system one would have the feeling that the gradient from 20^m to $18^m.5$ in the Draco system is much stronger; but beyond that I probably would not go.

Here, then, we have for the first time the color-magnitude diagram of one of the dwarf elliptical galaxies, and in general features it is what we should expect if these systems consist of pure Population II. For the Draco system it is impossible to derive the absolute magnitude and integrated color by any reasonable method, because it is so large and the foreground stars play such a role. But there is excellent evidence that these systems do not differ either in luminosity or in other properties from the Leo II system, for which we can determine the total magnitude and color index because it is in high galactic latitude and appears more condensed, being twice as far away. Holmberg has done this, using the preliminary color index, and the mean brightness of a number of cluster-type variables. In this way we find that systems of the Draco type have absolute magnitudes around -10, only slightly more luminous than the brightest globular clusters. Their color index on the International System is about $+0^m.8$; Holmberg's value for Leo II is $+0^m.83$. This value is in pretty good agreement with the mean value that has been derived for elliptical galaxies, which of course includes principally the elliptical galaxies of high luminosity on account of selection. According to Holmberg the mean color index for the elliptical galaxies is $+0^m.85 \pm 0^m.1$, a very small dispersion, based on about 15 cases. Stebbins and Whitford, from 36 elliptical galaxies, obtained $+0^m.87 \pm 0^m.01$. So in this respect the dwarf elliptical galaxies are representative of elliptical galaxies as a whole.

But today we know that this is not the whole story, on

account of Morgan's observations of the integrated spectra of a number of giant ellipticals. He studied giant ellipticals in the Virgo cluster. The line intensities, metal content, and so forth are quite difficult to interpret, because the lines of giant ellipticals appear very wide on account of the huge velocity dispersion of the stars within the system. If an elliptical galaxy contains 10^{10} solar masses, a velocity dispersion of the order of 200 km/sec would be expected from the virial theorem. The metallic lines are so much widened that they form blends, and we do not know whether these blends represent the same lines that are observed in normal spectra.

However, Morgan found a really significant result, that the cyanogen bands are quite strong in the spectra of these giant ellipticals. The cyanogen bands are always very weak in the spectra of pure Population II giants, so if the giant ellipticals are made up of Population II alone we should expect the cyanogen bands to be weak, just as they are in the globular clusters. But *de facto* Morgan found the cyanogen bands quite strong; perhaps (though he did not go so far as to say so) they are as strong as in normal giants.

As I mentioned in discussing the integrated spectra of globular clusters, for any reasonable luminosity function the integrated light is essentially determined by the brightest three or four magnitudes in such a system. Morgan's observation simply means, therefore, that most of the light in giant ellipticals comes, not from stars of low metal content, but from stars of considerably higher metal content. That these stars also must be old stars follows immediately from the same argument that I gave for the globular clusters of the disk: the giants must have come from somewhere relatively low on the main sequence and not from the region of A stars, because the integrated spectra of giant ellipticals are of types G to K, whereas if the brightest stars were young giants we should expect type A. So it is quite clear that these stars are old stars, but of higher metal content than the halo stars.

We must be very careful now, because it would be easy to jump to the conclusion that these giant stars are like those in M 67, which we know would do the trick. But we must really be a little more cautious, because Morgan does not make the statement that the strength of the cyanogen bands points to normal metal content. So I think we should keep an open mind about the possibility that there may be intermediate cases. Probably one way of deciding the question would be to investigate the disk globular clusters. Only from color-magnitude diagrams and detailed investigation of individual stars in such clusters can we show whether there is a whole range in between, or whether the M 67 case is realized. I suspect that intermediate cases are probably more likely, because we know that the strength of the cyanogen bands in the so-called high-velocity stars shows quite an erratic behavior; it is not just characteristic of one group, and apparently there is a large spread of chemical composition. Let us take the more open view that the evidence of the spectra shows that a large part of the light of giant ellipticals comes from enriched stars rather than from metal-poor stars; I think that is probably the fairest statement.

Is there any evidence that the Draco system contains two groups of stars, one of lower, one of higher metal content? I think the answer is definitely no. If there were two groups — even for the extreme case of M 67 — we should have a concentration of stars at 20^m, color index $+1^m.4$, falling off at fainter magnitudes, and I think there is no reason to suspect anything of this sort, because the stars in this part of the diagram are probably just field stars. If there were intermediate stars, they would be scattered all the way from 18^m to 20^m, and +1^m.2 to +1^m.4. So, as far as the diagram for the Draco system is concerned, there is certainly no strong indication that stars of higher metal content play any essential role. So in dwarf elliptical galaxies we seem to have a pure case of metal-poor stars, but in giant ellipticals there is such a mixture that most of the total light comes from enriched stars.

An intermediate case, where we can say something about the luminosity of the brightest stars that we observe, is NGC 205, the elliptical companion of the Andromeda Nebula. In this case we know positively that the brightest red giants that we observe are of absolute magnitude −3, just as in globular clusters; that is an observed fact. Does NGC 205 contain any admixture of stars of higher metal content? Morgan did not observe the spectrum, but Schwarzschild and Baum have investigated NGC 205 by the so-called count-ratio method in connection with the problem of the presence of stars of high metal content in the nucleus of M 31.

The count-ratio method works as follows. You assume a luminosity function for the pure Type II system that you are investigating. You can reach only the brightest stars in the system, and in each area of the system you count the number of these brightest stars. Knowing this number, you can now normalize the luminosity function for your area, and extrapolate for the number of fainter stars, and from this you can compute the surface brightness. If your system really has the luminosity function that you assumed, the observed surface brightness will agree with the one that you computed on the basis of the counts.

In this case we define the luminosity function by that

found for globular clusters. Of course there is a danger that a globular cluster is not a stable system; a considerable percentage of the faint stars may have been lost. But this is unimportant because only the bright stars play a role; 98 percent of the total luminosity is explained by summing up the first three or four magnitudes.

We can count the stars in an area in the outer part of M 31, and can then predict the surface brightness for this area by means of the luminosity function. Schwarzschild and Baum found a clear-cut discrepancy for M 31: the actual surface brightness is larger than that predicted. But for NGC 205 there was not such an obvious discrepancy, although the method is not very sensitive. We cannot state definitely that stars of normal metal content are not present; there may be such stars in NGC 205, but they do not play such a dominant role as they do in the giant ellipticals.

The absolute photographic magnitude of NGC 205 is −15, the dwarf ellipticals are around −10, and the giant ellipticals around −19 to −20, so NGC 205 is really a good case of intermediate luminosity. Without overstating the case, we can say that the metal-poor Population II is dominant in the dwarf elliptical galaxies, and that there is no indication of any important admixture of old stars of higher metal content. In the giant ellipticals, according to Morgan's results, the old stars of higher metal content are already dominant. For the intermediate case we cannot decide much, but there is no strong indication that metal-rich stars play any important role.

I think the prospects of getting a solution of the problem, by a new study of the integrated colors of these systems, are very bright. The integrated colors that have been derived up to now, photoelectrically by Stebbins and Whitford, and photographically by Holmberg, are for apparently bright elliptical galaxies, which means that they are for predominantly giant systems. So the mean color indices that we have at the present time are essentially those of giant ellipticals. But today we have such a large selection of elliptical galaxies in our neighborhood, over the whole range of luminosities, that we should make an attempt to see whether there is a correlation between integrated color and luminosity. I made such an attempt in 1946, after Stebbins and Whitford's first paper on the color indices of galaxies appeared. I took the distances of the galaxies simply from the velocity-distance relation, avoiding systems for which the red-shift was less than 2000 km/sec; beyond this distance you can determine the luminosity with a mean error of about $0^m.1$. I found a definite correlation between the color indices on the International System and the absolute magnitudes of the elliptical galaxies. These data were referred to the old distance scale, and if you introduce the new scale the difference would go up by a factor of two or three. I was especially interested to note that, for -8^M to -9^M, the color approached the observed color index of the globular clusters, of the order of $+0^m.56$ on the International System. Later on I became a little distrustful of the result, when I induced Stebbins and Whitford to measure a few more elliptical galaxies representing the lower end of the luminosity distribution.

In the Virgo system, M 100 is surrounded by quite a collection of dwarf elliptical galaxies, and, for some reason that I do not yet understand, two or three of them did not fit in very well. Finally, Holmberg derived a color index of $+0.^m83$ for the Leo system, and Baum checked this value with his photocell. The Leo II system, in contrast to the Draco system, is quite easily observed because it is about twice as distant, and is in latitude 79°. I gave Baum some original plates, so that he could select the best areas for photoelectric measures of the color index. Using these areas, he obtained the surprising result that the color index of the Leo II system is $+0^m.57$ instead of $+0^m.83$, and he has checked the measures on several nights.

Now this means that the Draco system and the Leo II system have not only the same color-magnitude diagram as the globular clusters, but also the same integrated color. Baum has measured a number of additional systems and taken data for others from the literature, and has plotted color index against absolute photographic magnitude, using the new distance scale. The most luminous galaxies have the high color index of $+0^m.85$ on the International System, and the color index levels off to about $+0^m.56$ for luminosities of about -11 or -10. The number of points is still very small, and the result depends on the correctness of these data. But I am now willing to believe that it will finally turn out to be this way, because the old results already indicated the same trend.

The globular clusters always presented the problem that their integrated color index was $+^m.56$, whereas the mean color index of the so-called pure Population II systems is $+0^m.85$. There were only two ways out. One could assume that, unlike the globular clusters, the giant systems contain a very much higher number of dwarfs; it could be done in that way, but whether it is the right explanation has always been doubtful. Now we see the entirely different possibility that, as the luminosity, and certainly also the stellar content, of elliptical galaxies increases, there may be an admixture of another sort of star, which produces this change in color index.

Unfortunately we cannot observe the brightest individual stars in one of the brightest ellipticals; even the nearest is too far away. There are two possibilities. One is that all the stars are richer in metals. The second is that there are two groups of stars, one of lower, the other of higher metal content. The latter possibility seems to be preferable at the present moment because, as we shall see in Chapter 18, we find this situation in the Andromeda Nebula, where most of the light in the central region is contributed not by the metal-poor stars of Population II, but by old stars of higher metal content. In this case we can observe the brightest stars, and we know they are metal-poor stars of absolute magnitude -3. So we know we actually have an admixture.

This idea of two kinds of stars has been suggested by Hoyle, Fowler, and the Burbidges, as a result of their work on the building up of elements in the stars. The first generation of stars, they say, is metal-poor; these stars eject part of their metals into space, and the second generation of stars can then build up the higher elements. Observationally the case in M 31 is very clear. The brightest stars are metal-poor, and you have to assume that this second group underlies the metal-poor stars. They must be old stars, but whether they are equally old we cannot decide with our present means.

But I believe that we can check the composition of the giant ellipticals without seeing their brightest stars. Many of them contain a large number of globular clusters; some have several hundred. If we could show that the color of these globular clusters is the same as that of the globular clusters in our own Galaxy, we could be sure that metal-poor stars were present. These clusters are representative of a larger population of the same type. So one should determine the colors of the globular clusters, say those associated with M 87 in the Virgo cluster. If they turn out to have the same color as the globular clusters of the halo, we shall have established the second possibility, that both metal-poor and metal-rich stars are present.

At the present time the form of the color-magnitude diagram is certainly a very much more sensitive indicator of metal-hydrogen ratio than any quantitative spectrum analysis that we can perform. So it is terribly important to compute intermediate evolutionary paths between the cases of the extreme globular clusters and M 67. I should rather like to keep an open mind about the whole problem, for the reason that the giant branches of the globular clusters that we know all lie in a very small area, although they range from M 3, where the difference from normal stars in the strength of the metal lines is hardly perceptible, to cases like M 92. It almost looks as though in this region the position of the giant branch may not be too sensitive to metal content, and then very suddenly it swings over. The only way to test it is by computing additional evolutionary paths for other hydrogen-metal ratios.

One further point about the elliptical galaxies concerns the emission lines of [O II] at 3727 and Hα, usually observed close to the nucleus. We find the emission in quite a high percentage of elliptical galaxies. It is of course not difficult to explain the emission, because among the Population II stars there are plenty on the horizontal branch that are able to excite it. Of course they will not be effective over such large distances as normal O and B stars, but they are certainly numerous enough to do the trick.

We have a number of data on the mass of gaseous matter at the center of these elliptical galaxies. The nearby systems like M 32 and NGC 205 have been observed at Leiden, and it turns out that the mass of gas must be less than 10^{-3}, and probably less than 10^{-4}, of their total masses. Osterbrock has recently studied a number of these systems, in the hope that the velocity dispersion for the gas (which is of course very much smaller than for the stars) will be small enough in some of them to permit the observation of the intensity ratio of the two components of the [O II] pair. He has found two cases in which the lines are separated, though they still partly overlap. From the density, and the area from which the emission comes, he concludes that the total mass of this gas in these systems is definitely less than 10^{-4} of their total mass.

It has been suggested from time to time that new star formation could take place from this residue of gas in the center of elliptical galaxies. This is most unlikely, because Osterbrock found that the gas concentrated in a small area in the center of most of these elliptical galaxies shows quite fast rotation. By turning the slit around, he could always find an area where rotation showed up. Gas has fallen in from greater heights, and it is not unreasonable to expect that it is not only in a state of high velocity as far as the atoms are concerned, but also rotating. There is probably no chance whatever of star formation from this gas. Its total mass is exceedingly small, so we can consider that these systems are practically gas-free.

Let me make a final remark on the diameter of systems like the Draco system. The distance modulus is $20^m.25$ for the Draco system, corresponding to a distance of 117.5 kpc. With an apparent diameter of 48' we get a linear diameter of about 1630 pc for the system. Such galaxies are probably the smallest that exist, and they are larger than the largest globular clusters by a factor of ten. For instance, the angular diameter of M 3 has been determined from its most distant cluster-type variables, of which it contains about 180. The diameters of the largest globular clusters are of the order of 100 pc, so it is clear that there is a big difference between elliptical galaxies and globular clusters. Certainly there is no reason whatsoever to believe that the globular clusters are the continuation of the elliptical galaxies. On the contrary, the fact that the mean mass-density of the globular clusters is so much greater than that of the elliptical galaxies simply indicates that they must have been generated in a system of lower density. If one group were a continuation of the other, we should find globular clusters with the same mass-density as these systems themselves. You have only to look at the Fornax system, or at one of the other dwarf elliptical galaxies in our neighborhood that contains globular clusters. You see right away that the globular clusters are intense knots.

It is remarkable that more than 50 percent of the members of our own Local Group are dwarf systems of the Sculptor type. The same is true of other nearby groups such as the M 81 group, and it is certainly true for larger groups like the Virgo cluster. This means that galaxies have a range of luminosity of 10 magnitudes, from about -20^M to about -10^M, and probably these dwarf systems represent more than half of all galaxies per volume of space. As far as masses are concerned, they probably do not play any role. If you want to compute the mass of our Local Group you have just to consider M 31 and our own Galaxy; you can forget about the rest.

CHAPTER 16

IRREGULAR GALAXIES

AND STAR FORMATION

THERE are two groups of irregular galaxies. One group, which we can call Irregular I, is rather blue, with a mean color index of $+0^m.28$ on the International System. The Irregular II systems have average color indices of $+0^m.84$, about the same as the integrated color of the elliptical galaxies. An example of the Irregular I systems is NGC 4449, a well-known Magellanic Cloud–type system outside the Local Group; its integrated spectral type is A7. We can take as examples of Irregular II the galaxies NGC 3034 and NGC 3037 (both companions of M 81) and also NGC 5363. The last of these has integrated spectral type G9, in close agreement with its color index, but NGC 3034 and NGC 3037 both have early spectral types in contrast to the color index; the spectral type of NGC 3034 is A5.

I should like to discuss an example of Irregular I galaxies, IC 1613, a dwarf galaxy belonging to the Local Group, on which I have worked for many years. It is a rather well-defined object of the Magellanic Cloud type. It shows two groups of very blue stars, one more conspicuous than the other, and there is also an indication of a kind of bar, which is a typical feature of most Magellanic Cloud–type galaxies.

The variable stars in IC 1613 were studied on plates taken with the 100-inch; 63 variables were found, and for 36 of these we have periods and light curves. The rest are essentially faint Cepheids, with periods of 2 or 3 days, and they lie so close to the plate limit that an investigation of their periods and light curves would hardly pay off. Of the 36 variables, 25 are Cepheids, seven are irregular, one is a long-period variable, one an eclipsing star, one a nova, and one a queer type.

The Cepheids range in period from $2^d.4$ to $146^d.35$. If one plots their magnitudes against log P in the normal way, they show a very small scatter around the period-luminosity relation; in fact there is just one star that deviates more than $0^m.3$ from the mean relation. Most of them lie within $0^m.25$ of it. All the standards have been determined photoelectrically on a strict photometric scale. If we use the new zero point of the period-luminosity relation (Shapley's original zero point increased by $1^m.5$) we obtain a distance modulus, $m - M$, of $24^m.1$ for IC 1613. Perhaps this value may later be changed by $0^m.1$. because the period-luminosity relation derived here does not quite agree with Shapley's in over-all slope, probably on account of some scale error in the latter. For stars with periods over 10 days the distance modulus is $24^m.13$, and for periods smaller than 10 days it is $24^m.07$. Until we have richer material for determining the slope of the period-luminosity relation we cannot rule out these small deviations.

Among the Cepheids is one with the very long period of

146 days, and the first thing to do was to decide that it really is a Cepheid and not a long-period variable. The amplitude is rather large, $2^m.16$, and the mean absolute photographic magnitude is $-5^M.40$; as far as the light curve is concerned, it has all the earmarks of a Cepheid. In order to settle the question I obtained two spectra when the variable was at maximum with the solid Schmidt spectrograph at the 100-inch; in both cases the star was within 10 days of maximum. Joy classified the spectrum as cK2: Ca I 4227 was very strong, and this line is normally not seen before K0, and the Sr II pair at 4215, 4077 were very strong, and that means high luminosity. Joy pointed out that the fact that the H and K lines of Ca II are exceedingly sharp points in the same direction. Humason classified the spectrum as somewhere between F5 and G on the basis of the strength of the hydrogen lines and H and K. Such an early type is of course to be expected if the spectrum is classified by means of the hydrogen lines, because it is well known that Cepheids have abnormally strong lines of hydrogen at maximum, so that the spectral type inferred from them is about 8 spectral subdivisions earlier than that derived from the metallic lines. So the star conforms to the behavior of a Cepheid in every respect; moreover, it fits the period-luminosity relation. The star is quite regular, except that in the early 1930's the minimum brightness changed by $0^m.2$ from one year to another.

With one exception, all the irregular variables are red. Their median photographic brightness is $-4^M.4 \pm 0.2$; they seem to be quite a definite group, because you encounter them around 19^m and 20^m, but at fainter magnitudes there is not a single one. The absolute magnitude $-4^M.4$ is very similar to what one finds for the red variables and all the M stars in h and χ Persei, so in both cases we are dealing with the same group. If these red variables were just a general phenomenon of the red giants, I should have found plenty of red irregulars among the fainter stars, but they are not there; the variables just fall in the small interval near $-4^M.4$.

One of the irregular variables is exceedingly blue, with mean absolute brightness $-4^M.0$ and maximum brightness $-4^M.4$. In general behavior, color index, and so on, this star is of the same sort as the blue irregular variables investigated by Hubble and Sandage in M 31 and M 33, which they thought formed a clearly defined group that could be used in the determination of the distances of galaxies. But they got for this group the very different absolute magnitude $-8^M.4$, four magnitudes brighter. I make this point because I think it throws a little doubt on the conclusion that this group of blue variables is very well defined, because here we find one that has the same slow variations, the same blue color index, but is four magnitudes fainter. I mention this as a warning that the group may extend to fainter absolute magnitudes, and, if there are only one or two such cases in a galaxy, it may be a little dangerous to base the distance on them. Anyhow, the matter should be further investigated in other systems.

The long-period variable has a period of 446^d and an amplitude of $1^m.8$. If it is a member of the system its photographic brightness at maximum is $-5^M.5$, and I think there is very little doubt that it is a member. I think — though I am not a specialist in these things — that the amplitude is rather small for a long-period variable; otherwise the star is very regular. Probably in the region of periods of 400^d to 500^d we have a mixture of sorts of long-period variables. Normally we should not expect them to have luminosities as high as $-5^M.5$, otherwise the Harvard investigations would have picked them up in larger numbers in the Magellanic Clouds.

The eclipsing system has a period of $3^d.775$ and maximum brightness $-4^M.3$; it has a well-marked secondary minimum.

The nova caught me unawares three years ago, when I was taking transfer plates for Selected Area 68 on two nights. On the first night it was not there; the transfers were taken with the 200-inch, exposure times 5 minutes, excellent seeing, and images can be seen down to 21^m. On the next night three plates were taken, and the nova was present at $17^m.5$, which means that its absolute magnitude was $-6^M.6$ at that time. As the three plates were separated by intervals of only 20 minutes, it is not possible to say whether the nova was rising in brightness or already going down; probably it was caught before maximum. This is interesting because it shows that novae will appear occasionally, even in these miserable systems.

The "queer variable" has a period of $28^d.687$, accurate to the third decimal, and is very red. The light curve is quite regular, and looks like an inverted β Lyrae curve. There is perhaps a little larger scatter around secondary maximum than elsewhere. The absolute magnitude at maximum is $-5^M.5$ and the star falls about a magnitude above the period-luminosity curve. I have never seen an animal of this sort.

The brightest blue star in IC 1613 is very easily determined, because the system is at galactic latitude $+60°$, where everything is very empty. It has apparent photographic magnitude $17^m.00$ and absolute magnitude $-7^M.1$, and it is of constant brightness. It is probably the brightest blue star on the main sequence. Let me add that even after a careful search of 200-inch plates I have not found a single star cluster in IC 1613.

I now turn to a question that has interested me very much for many years, the question of star formation in IC 1613, which I think is very illuminating. I have already pointed out the rather bright association of stars in the northeast corner of the system. When I had made the first comparison of red and blue plates in the 1930's, it turned out that practically all these stars were blue stars, and it puzzled me very much. All the blue stars in IC 1613 seem to be jammed into this corner of it. There is a second, smaller group, but it contains also a large number of red giants, whereas the number of red giants in the large group is very small. I simply could not understand how the Lord had managed to scramble all the blue stars of high luminosity into one corner.

The answer became quite clear when the first picture of IC 1613 was taken in Hα light. The region of blue stars is simply a superposition of H II rings. One is pretty well defined, but apparently it lies behind another, which contains plenty of dust, so that part of it lies behind and is absorbed. Of course you must imagine the whole thing staggered in space, one ring behind another. The inner region especially consists of a number of circular arcs, and immediately gives the impression that star formation has been going on in huge emission regions, which have finally overlapped completely. If you make counts, you find that the area near the center is still quite opaque. There are all kinds of emission regions. One of the brightest, by which the radial velocity of the system has been determined, has a diameter of about 17 pc. The largest has a diameter of 143 pc.

So star formation is going on in this whole region at the same time. There are O stars present, and all the gas in the region is now excited. The region over which the H II regions are scattered has a diameter of 460 pc. Later we shall see an even larger example in the Large Magellanic Cloud. I think that when star formation is going on in an area it spreads in some way like a disease; that is the definite impression that one gets.

I hope soon to obtain a color-magnitude diagram of this superassociation. We know already that it contains a very small number of red giants, which means that the association must still be very young. Probably the earliest of these H II regions are fading, because the excited stars are going down. I think the expansion of the bright H II regions might be measured; it should probably be tried. Simply from the superposition of all these arcs (seen in projection, of course) we have the feeling that the whole thing has expanded, and this tiny one will probably expand. If we apply the same principle to the Orion Nebula, we can conclude that most of the star formation has taken place. It is interspersed with gas. At present the most active spot in the constellation of Orion is the Orion Nebula itself, and there the formation of stars is going on with high intensity.

The other association in IC 1613 is of the same diameter or a little smaller. It must be older because it already contains quite a number of red supergiants. But the first association is the place where star formation is active now. It contains the long-period Cepheid (146^d), and it contains the star that I mentioned as the brightest star in the system. And the Cepheid of next longest period (40^d) is also in this association.

Now I shall turn to some other dwarf irregular systems, to show how local star formation, not necessarily over such a large area or on such a large scale as in IC 1613, can modify the whole picture. In some of these galaxies everything is relatively quiet at the present time; nothing is going on, and we see just old stars that were formed several billion years ago. Another is similar, although little nebulosities are just beginning to appear. Then we see one that looks almost as if someone had thrown ink over it: it shows H II regions which are probably not as far developed as in IC 1613. Another shows a number of emission nebulosities; they will expand, and afterward the stars will appear.

We must imagine that star formation does not go on with the same intensity throughout the whole system, but is localized, and goes on in jerks, so to speak. In NGC 2366 (which is probably not an irregular but an Sc) nothing special is going on at the present time in the main body, but there are little local areas where all the star formation is going on. When you look at any galaxy you must realize that star formation is going on temporarily as well as locally — in patches and in jerks.

If we keep this in mind, our whole conception of an irregular galaxy of the Magellanic Cloud type assumes an entirely different aspect. We should suspect, in many ways, that the ragged outlines of such a system are essentially determined by the region in which gas is still present, and in which star formation is still going on in jerks. We can show this beautifully for IC 1613, because we can easily pick up the Population II stars, far out, with the 200-inch. We find indeed that IC 1613 is very much larger than one would expect from photographs taken in the blue. On blue plates we observe an area of 17′ × 13′, but very ragged. But among the red giants of Population II, we observe a very regular elliptical area of 25′ × 20′, with the major axis lying in the east-west direction. The photographic picture lies eccentrically within it, with a ragged outline defining the position of most of the gas and consequently most of the star formation, and the old stars outside. So the irregularity simply indicates where star formation has been, and is still, going on. So the argument that irregular galaxies are chaotic, and that any chaotic galaxy must be young, rests on a misinterpretation of the facts.

CHAPTER 17

THE MAGELLANIC CLOUDS

T H E Magellanic Clouds are average-sized galaxies; the Large Cloud is already on the large side. They probably present the most magnificent example in our neighborhood of a collection of supergiant stars of all sorts. Although to talk about the Magellanic Clouds at Harvard is like bringing coals to Newcastle, they offer such interesting features bearing on stellar evolution that I should like to devote a chapter to them.

When the Large Cloud is photographed on a blue-sensitive plate, so that the supergiant stars are very conspicuous, one can see emission nebulosities scattered over the picture in large numbers. Every one of these emission nebulosities is associated with a cluster of very bright stars, which are obviously O stars, because they are very blue, and they excite these huge emission nebulosities.

If you measure the diameters of these groups of stars, you find that they are identical with what Ambartsumian has called associations in our own Galaxy; they run from 20 or 30 pc to about 100 pc. The emission nebulosities are scattered everywhere, and they are always accompanied by associations. The fact that the associations are always connected with gas is a very striking hint that stars have been formed in the gas. An association of large size is often near a very dense area, and (as has been checked at Pretoria) stars can be detected on large-scale photographs in these dense areas, so star formation is going on there. Moreover there are very few associations — none of them very bright — in which nebulosity is not visible on blue plates, and I am sure that it would be brought out by red-sensitive plates on account of the Hα emission.

A striking feature of the Large Cloud is the bar that forms its main axis; another is the Loop Nebula, 30 Doradus. Here also there are quite a number of what might be called distinct bright star clouds. We know that most of them consist of O and B stars; the *Henry Draper Catalogue* and *Extension* go faint enough to show that all the O and B stars are concentrated in these areas. Shapley noted them some years ago, and called them "constellations"; I think that by analogy with the term associations we can call them superassociations, because their diameters are of the order of 400 to 500 pc. We saw the same thing in IC 1613, but compared to these the superassociation in IC 1613 is a miserable affair; the number of stars here is 20 to 30 times greater. And every one of them stands out as a concentration of O and B stars.

You can also see at the first glance that the bright stars in these associations differ; some contain very bright stars, others fainter stars, so that you can pick out especially bright associations and less bright ones. Here we obviously see the effect of age differences. In many associations the upper part of the color-magnitude diagram has already disappeared, and there is one well-known association that still stands out, although it hardly contains any more O stars.

However, nothing of the sort is to be found in the bar, except in the large association around S Doradus. Here there are many emission nebulosities, and every one of them contains bright stars.

The biggest of the superassociations is one that is partly hidden by the bright nebulosities of the Loop Nebula; the whole area is filled with bright stars. Actually there is even a big cluster, which looks like a very young one.

I think that it is very important to recognize that star formation has apparently taken place on two scales — in the associations as defined by Ambartsumian, with diameters of the order of 10 to 100 pc, and over huge areas with diameters of the order of 500 pc, perhaps even 600 pc. One of the best illustrations is the Loop Nebula, which shows star formation going on over the whole area. These stars must of course be very young, for they are of the eleventh and twelfth magnitude, corresponding to -7^M to -8^M.

As there is star formation on two scales, smaller and larger, we must be very careful about the luminosity function in such systems. For example, nowhere in the bar do we find bright stars in such great numbers; actually there are some O stars in the bar, but very few of them. The O stars are concentrated in the associations, which means that no large-scale star formation has taken place in the main axis during the past 100,000,000 years or so; as far as star formation is concerned the bar has been rather quiescent.

A superassociation can probably remain visible only until it begins to merge into the background, until its brightest stars are comparable to those of the background.

This probably means that they merge into the background when they reach $-2^M.5$ or -3^M; unless the association has a very high density, they will then be lost in the general background. We must picture the luminosity function in such a system as a composite of all the different phases of star formation. We must be very careful with a luminosity function, as quite often determined, with all the data superimposed.

I think that the known superassociations should be exceedingly important in determining the true luminosity function, at least at the brighter end. The stars are very numerous, and we can attack the problem directly, so it is very much easier than for the stars in the neighborhood of the sun. Even if we knew only the upper limit of brightness in each association, I think we could extrapolate the color-magnitude diagrams and determine the approximate ages, and then we could determine a significant curve for star formation at the bright end of the color-magnitude diagram in the proper way.

Salpeter has done the same thing from the data in our own Galaxy, but it is very difficult to do, as it involves a dubious photographic luminosity function, a still more dubious photovisual luminosity function, and a derivation from a combination of the two. The proper way to determine the upper part of the luminosity function (or, if you want to put it otherwise, to determine the frequency of masses) is what I have described. The data from the Magellanic Clouds should amply suffice to do the thing down to 14^m or 15^m, that is from -5^M to -4^M. The curve could be extended to -1^M with material obtained with a large reflector.

Another remarkable thing about the Large Cloud is the

unbelievable number of star clusters that it contains. By contrast, not a single star cluster is found in IC 1613, although there is a large superassociation. Even if the upper limit of brightness goes down, a star cluster will remain visible for quite a long time, simply because of its high stellar density; it will always stand out as a fluctuation of density. So the fact that not a single star cluster is observed in IC 1613 means that no star cluster has been formed there for a very long time.

Per contra, the Large Cloud contains an amazing collection of star clusters, most striking to anyone who has worked on other galaxies. There are clusters of all varieties, from extremely rich ones down to small bits. And obviously, from the brightness of their brightest stars, they represent the whole age range; it is really a fantastic collection of data. I am sure that the star clusters of the Large Cloud can furnish the history of the whole evolution of at least one stellar system.

The number of clusters known at present is several hundred, but southern observers have pointed out that this is only a fraction of the number that can be reached with large telescopes. One can hardly look on any reflector plate without seeing half a dozen; it is amazing. We do not know why star clusters are so abundant in the Large Cloud; neither the Andromeda Nebula nor M 33 can compare with it. I think it would be exceedingly interesting to follow the history of star formation back by means of the clusters. They persist so long that we can recognize them even after the brightest stars are gone, simply because the stellar densities are so high compared to those in associations and superassociations. Even such a simple procedure as picking out the brightest star in every cluster that can be handled would give us a tremendous lead, and probably also a first insight into the age distribution of the clusters. And these things can easily be done today.

An even more amazing feature is that among the young clusters there are some that rival the globular clusters in richness. The young Population I clusters that we observe today in our own Galaxy, the Andromeda Nebula, and M 33 are always rather miserable affairs, but much confusion has been caused by the fact that the most luminous clusters in the Magellanic Clouds, both in richness and in luminosity, are of Population I. Among the most excellent examples is NGC 1866. They all resemble globular clusters in central condensation and in numbers of stars.

Until recently, no cluster-type variables were known in the Magellanic Clouds, and if there were no cluster-type variables there should be no globular clusters that showed a Population II diagram. In order to obtain information on this point, Thackeray examined the color-magnitude diagram of NGC 1866, by quite a simple method. The brightest stars in a globular cluster are red, and the brightest blue stars are $1^m.5$ fainter. In a Population I cluster the brightest blue stars should have about the same luminosity as the brightest red stars. The whole thing hinges on the colors of the brightest stars; if they are red, then the cluster is globular, but if there are mostly blue stars and a few red giants, then it is probably of Population I. It turned out that the brightest stars in NGC 1866 were actually blue, so it was clear that this is not a globular cluster. We also knew that NGC 1866 contains ten classical Cepheids, all with periods around 3 days, and since Cepheids of this period have absolute magnitude about $-2^M.5$ it followed that the brightest stars in NGC 1866 are of about -3^M.

Here, then, we have clusters that look exactly like

globular clusters, with exceedingly great concentration, exceedingly rich in stars, that belong to Population I. Even after Thackeray had carried out this test, he still maintained that NGC 1866 looked like a globular cluster. Well, "globular cluster" is a good Herschel term, but it tells nothing about the color-magnitude diagram. There is nothing comparable in our own Galaxy, not even the clusters that have at present the same luminosity, like h and χ Persei. By the time the brightest stars have disappeared, these will be very miserable clusters; their luminosity will drop very fast, because they do not contain very many stars, nothing to compare to NGC 1866. And the number of such clusters in the Magellanic Clouds is quite large; actually all the brightest are Population I clusters like NGC 1866. Even after their main sequences have disappeared down to about $+3^{M}.5$ they will still be very respectable globular clusters. It is remarkable that there is nothing of this sort in our own Galaxy, or in M 31 and M 33. Probably in the Magellanic Clouds we see clusters that are like what the globular clusters of our own Galaxy resembled at a certain stage. As NGC 1866 contains ten Cepheids, all of similar periods, it will probably show cluster-type variables 5 or 6 billion years from now.

It is quite simple to separate the two groups of star clusters in the Magellanic Clouds by means of their integrated colors. Gascoigne and Kron measured the integrated colors for about two dozen of the clusters in the Clouds, and found that they congregate about two maxima. The colors of the globular clusters, which show cluster-type variables in the Clouds, have integrated colors near $+0^{m}.58$, and those like NGC 1866 have a mean color index of $+0^{m}.15$. Up to now the two groups appear to be quite separate.

The number of star clusters in the Cloud is just fantastic, and it would be most revealing to pick out the brightest star that is definitely a member of each cluster. By detailed studies one could get the whole history of the system. I say the whole history, because we know that there are globular clusters there, which means that star formation started in the Clouds just as early as in our own Galaxy, something like 5 to 6 billion years ago.

I turn now to the Cepheids in the Clouds, on which so much work has been done at Harvard, and discuss the frequency distribution of their periods. The material is very rich; probably 3000 Cepheids are known in the two systems combined. It is quite clear from the period-frequency curves that, since all the Cepheids came from the main sequence, these curves must reflect the number of stars in the Clouds that came from a given interval of absolute magnitude: each period is determined by an absolute magnitude. In the simplest case, we can assume that every star that left the main sequence has to pass through the Cepheid stage. Then the period-frequency distribution would represent the frequency distribution of the absolute luminosities where they turned off the main sequence. In the less simple case, if only a certain fraction of these stars have become Cepheids, this fraction must still be related directly to the total number. The fact that there are differences in the period-frequency relation for the two Clouds has certainly nothing to do with physical differences between the Clouds; it simply shows how many stars are in the Cepheid stage. We have seen that the numbers of stars are determined by the rate of star formation, which goes on in jerks. I think that the difference has nothing to do with physical state, but simply reflects the number of stars in each stage. If there are differences between the

Clouds, it simply means that there are differences of absolute luminosity, which are known to happen and are nothing to get excited about.

It has been suggested that the Cepheids of our own Galaxy differ from those in the Clouds because the Cloud Cepheids are bluer than those in the Galaxy. Today it is pretty clear that the whole trouble lay in our lack of knowledge of the intrinsic colors of the Cepheids in our Galaxy. The whole problem of the zero point of the period-luminosity relation lay, not in the proper motions, which were accurate enough, but in our lack of knowledge about absorption and reddening, which we always underestimated. I think that this is still the case. In fact, I believe that the measures made by the Leiden people in South Africa need only a final adjustment to bring everything into agreement.

Arguments have also been based on the fact that the amplitudes of Cepheids determined in the Large Cloud, the Small Cloud, and the Galaxy are different. I give no weight to this argument, because it is notoriously difficult to determine decent amplitudes.

The type of light curve as a function of period is very well observed. It is touch and go whether there are any differences between the two Clouds and the Galaxy. In all these systems, and also in M 31, as we shall see in the next chapter, the relation of light curve to period is followed as well as we can reasonably expect.

For the stars of shortest period there is a problem. In this region of periods a little longer than 1 day we still have trouble in disentangling the Type I and Type II Cepheids in our own system. Also we do not yet know at what period the Type I Cepheids stop, and whether they peter out. The frequency maximum for the Small Cloud still seems to lie at somewhat shorter periods than for the Large Cloud, but I think we are still dealing with Type I Cepheids, because I believe that Population II is too weak in both Clouds to make itself felt. But we do not know.

Regarding the old Population II stars in the two Clouds, I can be very brief. When Wesselink and Thackeray found cluster-type variables in some of the clusters, the presence of globular clusters of Population II was finally clinched. Recently cluster-type variables of the same brightness have been found outside the globular clusters, the first indication of field stars of Population II.

Today we have well-determined photometric contour lines for the Small Cloud. The star counts down to 16^m agree very well with straight photometric measures made with the photocell. It has also been found that there is a distinct color change as we go out from the central region of the Small Cloud. The tendency is general, not systematic: the outer regions have larger color indices than the central region. Elsasser has recently determined the integrated color by drawing photographic and photovisual contour lines; the integrated color index is $+0^m.14$ on the International System, which agrees quite well with the mean values given in the previous chapter. But it is interesting that for the outermost contours he gets the high color index $+0^m.87$, of the same order as in the elliptical galaxies of Population II. Hogg's photoelectric data, which do not go quite so far out, confirm the result. This means that in the outermost regions of the Small Cloud we have essentially pure Population II, where no star formation is going on at the present time, just as in IC 1613, where these red stars have actually been observed.

These red stars are probably not observable in the outer regions of the Small Cloud, but there are globular clusters, and Shapley found Cepheids in these outer regions, which (as he emphasized) have very short periods, around 2 days. It is quite possible that a mixture of populations is coming in; here it is not too difficult to pick up Population II Cepheids, whereas in the denser parts of the Clouds they are drowned out by the rich background, or at least they are very much more difficult to observe.

It would be very difficult to distinguish the Population I and Population II Cepheids; I cannot suggest any criterion by which we could tell them apart by light curve or by color index, unless we knew the difference of luminosity. It would be really important to find a case in which we could be sure that we were dealing with one case or the other. A region that I think should be very carefully searched down to the faintest stars is the Carina region. It is obviously very rich in Population I objects, and there one might get, to a first approximation, a reasonable estimate of the number of Cepheids with periods around 2 days. But it would only be an approximation.

Finally I will say a few words about the most luminous stars in the Magellanic Clouds. An attempt has been made at Pretoria in recent years to pick them out. The O and B stars are easy, for they are available from the *Henry Draper Catalogue* and the *Henry Draper Extension*. The difficulties start with the later types, because in these latitudes the number of foreground stars is very high. Radial velocities should be determined by objective-prism methods, with which Fehrenbach has had the first real success. It would be easy to sort out the late-type stars with an instrument like Fehrenbach's, because the radial velocity of the Large Cloud, for instance, is −240 km/sec.

Even now the results are interesting enough. When absolute magnitude was plotted against spectral type, it was found at Pretoria that there are no O stars brighter than -7^M, where the main sequence apparently reaches its topmost point. From there upward stretches a series of stars which reach almost -10^M at type Ao, a rather well-defined upper limit. Beyond Ao there is a gap (which may be spurious), and then three or four stars of types F and G. Colors have been determined for these stars, but they are not quite certain, because practically all these luminous stars show intrinsic variability; but in general the colors are in very good agreement. These bright stars will be especially interesting in connection with the upper limit of stellar masses. There are considerable theoretical difficulties in determining how large the mass of a star can become. I believe that the mass-luminosity relation gives not much more than 80 to 100 solar masses for a star of absolute magnitude -10^M.

CHAPTER 18

THE ANDROMEDA NEBULA:

PHOTOMETRY

IN CHAPTER 5 we saw that the Andromeda Nebula is a system where dust and gas are strongly concentrated to the arms, and that absorption produces very large effects in the inner arms, enhanced of course by the inclination of the system to the line of sight. Spiral arms can be traced to within about a degree of the center, but the emission nebulosities there are so heavily obscured and reddened that they show up only on red-sensitive plates.

More recently I have studied the variable stars in the Andromeda Nebula. Four fields were chosen for investigation. Field I takes in the central region, including a spiral arm which is indicated by a few patches of absorption. Field II covers a large part of this dust arm, which contains the huge star cloud or superassociation NGC 206. This is a real superassociation, one of the largest we know at the present time, with a diameter of the order of 1100 pc; apparently star formation took place nearly simultaneously over this

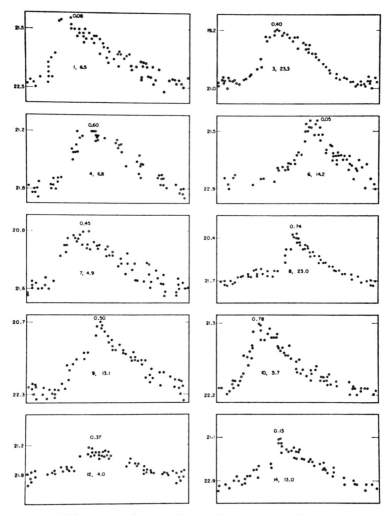

FIG. 26a. Light curves of Cepheid variables in M 31. Below each curve is the serial number of the variable, and the period.

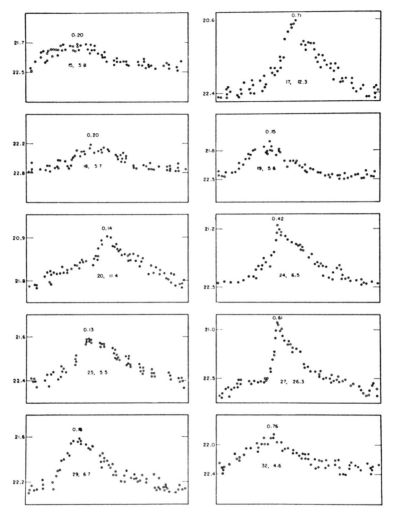

FIG. 26b. Light curves of Cepheid variables in M 31. Below each curve is the serial number of the variable, and the period.

whole area. Field III was also the main field studied by Hubble for variable stars; it is distinguished by being very clear, with less absorption than in any other part of this same spiral arm. The arm contains a star cloud, and then goes over to a place where it is essentially visible as absorption; but in the region studied it is especially clear. In Field IV we avoid absorption as much as possible, and we see hardly any spiral structure — just a little knot. These investigations were begun with the 200-inch as soon as it went into operation.

The number of variables found in the four fields is shown in Table 14. The material is not yet quite homo-

Table 14.   Variable stars in the Andromeda Nebula.

Field	Distance from center (')	No. of variables
I	15	150
II	35	223
III	50	366
IV	96	50

geneous. In Fields I and IV the search must be complete. The least complete is probably Field III; I am sure that if it were searched as completely as Fields I and IV the number of variables would be close to 500. The number of variables found is of course only a small fraction of the total for the Andromeda Nebula. A complete search of the whole system would probably end up with something like 8000 to 10,000 variables that are within reach with our present means.

The light curves for a number of variables in Field II are shown in Fig. 26. Eclipsing stars are quite common, and have been found in all the fields. Most of the work on Fields I, III, and IV has been done by Miss Swope and myself; Sergei Gaposchkin worked on the variables in Field II.

Besides eclipsing and irregular variables we found a large number of stars where one can simply say that they are faint and of short period; many of them must be Cepheids with periods shorter than 3 days; the number gets very large and it does not pay to work them all up. Finally there are the long-period variables, which are important. Their color at maximum, when we can get it, is very red, around $+1^m.5$. In half a dozen cases we were able to determine a period, especially in Field III, which has been observed through the years. From a number of maxima, it turned out that the periods are between 150^d and 250^d, so these stars fit in with the long-period variables observed in globular clusters. Twelve such stars give a mean magnitude of $21^m.85$ at maximum, which, with the distance modulus I shall give you later, leads to about $-2^M.4$. These values give an upper limit, because they come within a magnitude of the plate limit, about $22^m.8$ in this series with the 200-inch. It is quite certain that we have a selection of just the brightest long-period variables. If we could actually have observed uniformly down to the plate limit, the mean would probably have been $0^m.3$ to $0^m.5$ fainter, so $2^M.4$ is undoubtedly a little too bright. For long-period variables with periods between 150^d and 250^d in globular clusters we found a maximum magnitude somewhere around $-1^M.9$. So we need not worry about a discrepancy; it is quite clear that there is a selection effect. The stars that get only $0^m.2$ above the plate limit have been dropped out, and they would pull the mean down considerably.

The distribution of type of variable for two of the fields is given in Table 15. In Field I, a very difficult field close

Table 15. Distribution of types of variables.

Type	Field I		Field III	
	No.	Percent	No.	Percent
Cepheids	38	32	237	70
Short	30	24	50	14
Eclipsing	5	4	30	9
Irregular	32	27	20	6
Long-period	8	7	2	1
Novae	7	6	1	0
Total	120		340	

to the nucleus, 32 percent of the variables are Cepheids. The "shorts" are faint Cepheids for which no light curves have been derived. The number of irregular variables is quite high, and seven novae were found; the field is close to the nucleus, and novae show up in large numbers. In Field III the Cepheids dominate, and Sergei Gaposchkin's results run essentially along the same lines, for Fields II and III do not differ much in their relative distribution. One nova was found in Field III, of the Nova Herculis or T Aurigae type, with a long-lasting maximum for about 100^d.

The period-luminosity relation for Field II shows a huge scatter, the effect of absorption in its fullest display, which of course was to be expected. Field III was the best one could find; the absorption in the inner spiral arms is exceedingly strong.

That the scatter is really connected with absorption is shown by the approximate period-color relation. The blue stars lie at the top, then come the intermediate stars, and the red stars lie below. The stars of smallest amplitude are those that are bluer than one would expect from their periods. In every case the reference is to color at maximum. The most important feature is the large dispersion of the period-luminosity relation, caused by absorption in the spiral arms. It is astonishing that Hubble did not notice this large dispersion, because he worked in the same area; but we must not forget that his material was very limited compared to the present. All the periods that he determined were verified within $0^d.1$.

Neither the Population II Cepheids nor the long-period variables show any relation to the spiral arms; they are scattered all over. But the Cepheids always show a preference for the spiral arms.

If we form a scatter diagram and avoid the Type II Cepheids, we find that the mean deviation of a single point from the mean line is about $0^m.82$, and we can use this value to get an approximate value for the total absorption in Field III. Let us assume that the scatter is distributed along a Gaussian function. The $0^m.82$, the largest deviation of one point in one direction or the other, should be three times σ_{tot}, the total dispersion, and σ_{tot} should be $0^m.27$. But σ_{tot} is composed of two parts, namely σ_{abs}, the scatter due to absorption, and σ_{int}, the intrinsic scatter, so $\sigma^2_{tot} = \sigma^2_{abs} + \sigma^2_{int}$. We do not quite know the value of σ^2_{int}, but we can get an approximation by using the data from IC 1613 and the Small Magellanic Cloud, because in both cases we have reason to believe that the absorption is quite small. From the best light curves which Shapley published we get $0^m.17$ for the Small Cloud, and I get $0^m.15$ from IC 1613, for σ_{int}. With these values we get $\sigma_{abs} = 0^m.22$, so the scatter due to absorption is larger than the intrinsic scatter. But keep in mind that the figure used for the intrinsic scatter is still too large, for there is certainly some absorption in the Small

Cloud as well as in IC 1613. We have thus obtained a rough idea of the dispersion caused by absorption alone.

To get an idea of the total absorption in the main plane of the Andromeda Nebula, we must make some assumption about the distribution of absorbing matter there. In our Galaxy, the best thing we can do at the present time, until we have color data for each Cepheid, and can treat each one separately, is to assume that the density of the absorbing medium has a distribution like $D = D_0 e^{-2/\beta}$. According to the Leiden measures, the neutral hydrogen in our Galaxy is distributed according to a formula of this type. Schmidt has shown, for instance, that for the interstellar medium in our own Galaxy, or for the neutral hydrogen, we can write this as:

$$D = D_0 \times 10^{-0.0038|z|}.$$

There is good reason to believe that the neutral hydrogen in M 31 would be very similarly distributed to that in our own Galaxy, according to a formula of this type.

We then get the following result. We view the plane of the Andromeda Nebula at a small angle. It is easy to see that the absorption, for any point at a distance z from the plane, will be

$$A = \bar{A} \times 10^{-0.0038|z|}$$

for points lying above the main plane, and

$$A = 2\bar{A} - \bar{A} \times 10^{-0.0038|z|}$$

for points lying below the main plane. We expect that the absorption is symmetrically distributed about the plane. For a point lying in the plane itself, for $z = 0$, we get the absorption $\bar{A}$. Schmidt's figure for the dispersion in z is 110 pc. If $\sigma_{abs} = 0^m.22$ and $z = 110$ pc, we obtain $\bar{A} = 0^m.58$, which is the total absorption in the plane. Thus for any Cepheid lying in the plane in Field III the total absorption is $0^m.58$. I mentioned that Field III is one of the clearest regions. We can write $\bar{A} = A_{90} \csc i$, where i is the angle at which we view the plane, and A_{90} is the absorption we would get if we viewed it perpendicularly. Since i is about $11°.7$, $\csc i = 5$; thus $A_{90} = 0^m.58/5 = 0^m.12$. So if we viewed Field III perpendicularly, the total absorption would be $0^m.12$, half as much as in the neighborhood of our sun. So Field III is actually clearer than the region surrounding the sun.

We now have the figure that we need to derive the distance modulus of the Andromeda Nebula. I have taken the cross section of the period-luminosity relation at log $P = 0.80$, or $P = 6^d.31$, and determined the mean magnitude here. From the diagram we find that the median magnitude at this period is $21^m.87$, and we have to introduce an absorption of $-0^m.58$, which gives $21^m.29$ for the corrected median. Since, from the new period-luminosity relation, the absolute magnitude at log $P = 0.80$ is $-3^M.03$, we get the distance modulus, corrected for absorption:

$$m - M = 24^m.32.$$

That is as much as we can do at present in this rough way. It is a makeshift, and the only thing that speaks in its favor is that it is quite reasonable.

There is another way to derive the distance modulus. We can determine the magnitudes of the brightest stars of Population II, say in the region between the Andromeda Nebula and NGC 205, where the stars are not yet mixed with those of NGC 205. The magnitude of these brightest stars of Population II is found to be $22^m.75$. This rests on a large number of transfers of Selected Area 68. Baum, who has measured the same thing photoelectrically, though not yet with the highest precision, finds for an upper limit the same value, $22^m.7$. These are the brightest stars of Population II outside the main body, and in a region where absorption in M 31 certainly plays no role. These stars have absolute magnitude $-1^M.50$, as we know from globular clusters. If we take the upper value, we find from these stars, for the distance modulus of M 31:

$$m - M = 24^m.25.$$

Now this is not an independent determination at all. We have adjusted our zero point in such a way that the absolute magnitude of the cluster-type variables was $0^m.0$, and have assumed that the cluster-type variables were correct. All the other zero points are related to this one. The whole thing simply shows that we are on the same magnitude scale. The two values of the modulus agree closely; with the data we have, they *should* agree to within $0^m.1$.

One thing is certain: we must consider this distance modulus as an upper limit. We assumed that there was no absorption in the Small Cloud and in IC 1613, and threw everything there into the intrinsic dispersion. Furthermore, we have taken no account whatsoever of the absorption arising in our own Galaxy in the line of sight between us and these Cepheids. We know in general today that the absorption in this direction must be small. In Field IV, where even the photoelectric people can work quite easily, the blue stars are practically as blue as anything in our Galaxy, so that they seem to have very little color excess.

We have good reason to believe that this upper limit may very well go down. The true distance modulus may be smaller by $0^m.1$ or $0^m.2$, perhaps even $0^m.3$. The data that are uncertain have an uncertainty of the order of $0^m.1$. So, even as we see the problem now, the distance modulus may perhaps go down to $24^m.0$. But one thing is certain: recent proposals to make the distance modulus of M 31 as great as $24^m.6$ or 24.7 are entirely impossible. That would require, in spite of the huge dispersion that is observed, that there is no absorption, and that everything is intrinsic dispersion. Everything speaks against it. So it is impossible to increase the distance modulus of M 31; it would be a complete retrogression. The arms of the Andromeda Nebula are loaded with dust and gas, and Field III is one of the clearest spots that we can find. A distance modulus of $24^m.6$ or $24^m.7$ is really out, and I am willing to bet with anybody.

In my opinion — and I am thoroughly convinced of it — the best way to determine the distance modulus of the Andromeda Nebula is via the brightest Population II stars in the outer parts, where one does not have to fear any absorption. One should, of course, work with photovisual magnitudes; the only thing that would be necessary would be for our photoelectric observers to determine a sequence, so that we could determine the upper limit of luminosity of the Population II stars as sharply as possible. As soon as you take Population I stars you run into the awful problem of absorption; because csc $i = 5$, you get this extra factor of 5 all the time.

Now let us consider the total integrated absolute photographic magnitude of the Andromeda Nebula, $4^m.33$ according to Holmberg. That makes the absolute magnitude -20.0 photographically, -20.8 photovisually. This is independent of the absorption, at least to the first order. Holmberg has calculated how much brighter it would be if it were viewed face-on, and it turns out that it would have to be increased by practically a magnitude to -21^M.o. The faintest systems, like Sculptor, Leo II, and Draco, are about -11^M, so there is a range of 10 magnitudes in luminosity.

If we were to put M 31 into the Virgo cluster or the Coma cluster it would be one of the brightest members, especially if we included the correction for turning it face-on. It is one of the brightest galaxies that we know. There are one or two galaxies in the Humason-Mayall list of red shifts that may be half a magnitude brighter, but these values are still uncertain.

For the first time we are in a position to make a comparison with the size of our own Galaxy. There have been many earlier attempts to compare the two, for instance by determining how far out the cluster-type variables of our Galaxy extend, but the thing never worked very well simply because we cannot observe the cluster-type variables in M 31. But for the first time I think we can do it today, since the spiral structure of our own Galaxy has been analyzed, simply by comparing the dimensions of the most distant arms that we can see in both systems. Of course we have the difficulty that the distance modulus of M 31 is an upper limit; I assume $24^m.1$ as the probable true distance modulus, corrected for absorption, especially absorption in our Galaxy. The most distant arms on M 31 are 21.9 kpc from the center. If we use the Leiden maps for our Galaxy, taking the mean for the four or five most outlying features, we find 13.6 kpc. So in linear dimension the ratio of the Andromeda Nebula to our Galaxy is 1.61; there is no doubt that the Andromeda Nebula is 50 percent larger than our own system.

Now consider the volume of M 31; it will be $(1.61)^3$, or four times, the volume of our Galaxy. The new determinations of the masses of the Andromeda Nebula and of our own Galaxy, based on Schmidt's models, lead to a ratio of masses about 4:1. This would mean that the mean mass density is roughly the same, rather a satisfactory result, which shows that we are not light-years away from the true solution..

For the globular clusters I use the data published a number of years ago by Seyfert and Nassau. This work, based on the North Polar Sequence, has meanwhile been shown by Mayall and Kron to be strictly on the International System. I have taken Seyfert and Nassau's list, because it is larger than that studied by Mayall and Kron. Table 16 shows the beginning of the general luminosity

Table 16.   Magnitude and number
of globular clusters in M 31.

Mean apparent magnitude (m)	Number of clusters
14.75	0
15.05	2
15.35	2
15.65	6
15.95	10
16.25	15

curve for the globular clusters in the Andromeda Nebula; it gives the number of globular clusters in an interval of $0^m.3$ centered on the apparent magnitude given in the first column.

If we draw a smooth curve we find the last globular cluster on the bright side at about $14^m.9$. To convert to absolute magnitudes we can use our modulus directly, both sets of data being equally affected by absorption. The brightest globular cluster then comes out at $-9^M.4$, which agrees quite well with the data for our Galaxy; ω Centauri is about $-9^M.3$.

Let us now take the beginning of the luminosity function for the brightest stars in M 31, again using the data of Seyfert and Nassau, because they have made the largest investigation so far, covering the whole Nebula (Table 17). The number of stars is tabulated for half-magnitude

Table 17. Magnitude and number
of luminous stars in M 31.

Mean absolute magnitude (M)	Number of stars
-8.95	3
-8.45	9
-7.95	24
-7.45	59
-6.95	183
-6.45	440

intervals. The absolute magnitudes again are based on a modulus $24^m.32$, since the absorption cancels out. We see that there are certainly stars as bright as -9^M in M 31, and the numbers given here are definitely too small, because these most luminous stars will all be lying in the spiral arms, and therefore be heavily affected by obscuration, so that they are pushed down in the list toward fainter magnitudes. Certainly the upper limit in M 31 goes up to -9^M.

The novae in M 31 give similar results. Arp's reinvestigation gave really good light curves and reliable magnitudes. He obtained $-7^M.45$ for the mean magnitude at maximum, and the best determination for our own Galaxy, by Cecchini and Gratton, has $-7^M.4$ or $-7^M.5$.

Although the agreement for our Galaxy is so good, there is evidence of real differences. For example, if you examine the luminosity function for the globular clusters, as known at the present time, you find two peaks, one near -8^M and one near -6^M. We know that the latter is underestimated at the present time, because it is very difficult to measure the integrated magnitudes of very faint blue globular clusters, but even after the Schmidt survey on Palomar I doubt whether it will be as high as the other peak. In M 31, on the other hand, the peak for lower absolute magnitudes is much higher, so there are real differences. But fortunately the upper limit of absolute magnitude is the same in both systems. So far as we know, the Andromeda Nebula is just as old as our Galaxy, so the difference cannot be a result of difference in age.

I will conclude with some remarks about the sheet of Population II stars that are brought out in M 31 with the 200-inch telescope. Simply from the upper limit of brightness, we know that these are the brightest giants of the metal-poor Population II. But, as Schwarzschild and Baum first pointed out, the light emitted by these stars is not enough to account for the total surface brightness of the Andromeda Nebula. They counted the brightest stars of Population II that they could see down to a certain limit, and, using Sandage's luminosity function for M 3, they used these data to compute the surface brightness. They found that the measured surface brightness was considerably greater than this. However, they did not prove that the stars responsible for the difference were ordinary giants; they only showed that something else must be present.

Morgan has clarified the matter more recently in a very simple and direct way, by putting the slit across the central region of M 31 and obtaining the integrated spectrum. In this case, just as in his comparison of the globular clusters of the disk and the halo, one cannot use the intensity of the absorption lines, because one can see the great width of the lines even with moderate dispersion of 100 or 150 A/mm, and the lines are of course blended. It is not possible to reach conclusions about details of spectral type with these data. Minkowski's unpublished work shows, in fact, that one has to use rather large dispersion, up to 20 A/mm, to find iron lines that are not blended, and these lines are too faint to show up on Morgan's small-dispersion spectra.

However, Morgan's spectra, taken at McDonald Observatory with a quartz spectrograph, showed that the cyanogen bands in the ultraviolet are very strong in the integrated spectrum — quite as strong as in normal giants. Now Morgan and Keenan had already shown that one of the differences between Population I and Population II stars is that metal-poor stars have very weak cyanogen bands: in general the strength of the cyanogen increases from dwarfs to giants to supergiants, but the Population II stars form an exception. De facto, on the basis of the cyanogen bands, one would classify Arcturus, which is metal-poor, not as a giant but as a dwarf. In their *Atlas*, Morgan and Keenan pointed out that one has to be very careful with the cyanogen criterion, because it will lead to wrong classifications with high-velocity stars of low metal content.

In the nuclear region of the Andromeda Nebula, Morgan found the cyanogen bands of the integrated spectrum just as strong as in normal G or K giants: he classified the spectrum as somewhere between late G and early K with normal cyanogen — not very precisely, because of the width of the lines. He concluded that most of the light that we receive from the central region of the Andromeda Nebula comes, not from the dense sheet of Population II giants, but from underlying stars of lower luminosity, which yet produce most of the light. So he concluded (and I think it is a very direct and beautiful proof) that most of the light comes from what he calls CN giants, ordinary giant stars.

Even these CN giants must be old stars; they cannot be young. We know that the ordinary giants of absolute magnitude 0^M or -1^M are fed into the giant region by two routes: one from a position near type A on the main sequence (the Hyades-Pleaides stage), and one from a position represented by a system like M 67. I do not say that it is exactly M 67, but it is similar. Morgan's classification of the integrated spectrum excludes the turnoff at type A, because in that case the main sequence would be strongly represented by A stars and the composite spectrum would be of early type, A5 or A8 or F, certainly not late G or early K. This points to something like the M 67 case as the solution.

Minkowski is working with the four unblended iron lines to find the velocity dispersion in the nuclear region, and also the spectral type, and he finds a giant K0 spectrum from these four lines, rather than the dwarf G5 formerly derived by Humason and Adams. So in any case the main sequence has largely disappeared up to $+3^M$ or $+3^M.5$. The main point is that the stars that produce the cyanogen bands must be old, though not necessarily as old as the stars in the galactic halo.

We have the same problem in the big elliptical systems, where, as Morgan has pointed out, it is practically impossible to determine an exact spectral type from the widened absorption lines. For the spectral type we have to fall back on the cyanogen bands.

The large velocity dispersion is a result of the large mass concentration at the center. The only way of determining the central mass is via the velocity dispersion; one uses the virial theorem for the inner part. One finds right away that the velocity dispersion for the stars should be several hundred kilometers per second. The gas, on the other hand, has a very small velocity dispersion, as Spitzer pointed out many years ago.

We must conclude, then, that in the central region of the Andromeda Nebula we have a metal-poor Population II, which reaches -3^M for the brightest stars, and that underlying it there is a very much denser sheet of old stars, probably something like those in M 67 or NGC 752. We can be certain that these are enriched stars, because the cyanogen bands are strong, and so the metal/hydrogen ratio is very much closer to what we observe in the sun and in the present interstellar medium than to what is observed for Population II. And the process of enrichment probably has taken very little time. After the first generation of stars has been formed, we can hardly speak of a "generation," because the enrichment takes place so soon, and there is probably very little time difference. So the CN giants that contribute most of the light in the nuclear region of the Nebula must also be called old stars; they are not young.

Now Morgan only investigated the central region, and the density falls off so rapidly from the center that it would be very difficult to get uncontaminated spectra for regions farther out, because we must always fear that the contribution of skylight will play havoc with the lines that show up. But, as I have already mentioned, Schwarzschild and

Baum measured surface brightness and color index for a number of areas on the outskirts of the Andromeda Nebula, for instance between the Nebula itself and NGC 205; these regions are far outside the spiral structure. They found the remarkable result that the integrated color indices for these outlying areas were definitely lower than +0^m.87, the value for the nucleus. Between the Nebula and NGC 205 they found values as low as +0^m.74.

When I discussed the Draco system in Chapter 15, I pointed out that, according to its color-magnitude diagram, this system is metal-poor, just like the globular clusters, whereas the giant elliptical M 87 is a system of just the same sort as the Andromeda Nebula, because the cyanogen bands show up strongly. There are plenty of data for giant elliptical galaxies, and the mean color index is +0^m.87 with a very small dispersion. Contradictory values were found for the color index of the Draco system. Holmberg's value was close to +0^m.87, but the method by which he obtained it, from integration on the photographic plate, must have been an exceedingly difficult one. Baum, on the other hand, found a value close to +0^m.58, as in the globular clusters. If this is correct, it means that the Draco system is a dwarf system with a pure halo population, while as we go to richer and richer systems, up to the giant elliptical galaxies, the admixture of CN giants comes in more and more and more and more.

We know the results for the Andromeda Nebula: the color index in the center is +0^m.87, and farther out it is +0^m.74. This would indicate that, as we go out from the center, the contribution of the CN giants becomes less and less. This is a point that should be very carefully investigated. The data for M 31 are quite sure, but they should be extended to one of the giant elliptical galaxies like M 87, to see whether there is indeed a change in integrated color toward the outside. We shall return to this point when we discuss the nucleus of our own Galaxy (Chapter 21), because of Morgan's recent investigation of the spectrum there. The situation is not so simple as in the Andromeda Nebula.

CHAPTER 19

EVOLUTION OF GALAXIES

I SHOULD now like to summarize the main points that we have discussed regarding galaxies, and draw some conclusions. We saw at the beginning that distribution of galaxies in space is very far from uniform. A uniform distribution would be practically no approximation at all, for we know that galaxies occur in doubles, triples, quadruples, groups; after the groups come clusters, and after the clusters, huge clouds of galaxies, such as the great cloud found by Shane.

On the other hand, we can classify all the galaxies that we observe by a very simple scheme such as Hubble's, especially if we restrict the group of irregular galaxies strictly to the Magellanic Cloud type. Hubble's system is really very much better than it looks from his own classifications, because in many cases he did not recognize systems that are physically double, and always classed them as irregulars, so that irregular became a catchall for systems whose nature was not recognized. I know only half a dozen galaxies, just a handful, where one has to press a little in order to fit them into the system, and even in these cases one can still say whether the system belongs in the spiral group or the other. I think the fact that it is possible to classify the forms of galaxies on such a very simple scheme, leaving out all little details that are not important, is very remarkable.

After the 200-inch came into operation I became very much interested in finding out whether we could find any galaxies that did not fit into the Hubble scheme. It is easy to take a plate that shows 2000 distant galaxies. During the past six years I have used practically every occasion of excellent seeing to photograph fields of galaxies in the blue and in the red, and to see whether I could find anything abnormal by intercomparing them. The astonishing thing is that I never found an exceptional case, and I carried the study down to galaxies with average diameters of probably 5″.

If you go to a diameter of 5″ we can say that you go to a distance of about 10^9 light years. For detailed study, of course, we have to fall back on the Local Group of galaxies, and there we are in a fortunate position, for the Local Group contains essentially the same types as we find in the general field. De facto, the only important type missing from our immediate Local Group is the barred spiral, but even that is not so bad, for there is a magnificent barred spiral in the NGC 2503 group that is beautifully resolved with the 200-inch. So we are fortunate that all the types are represented in the Local Group, or else we can fall back on one of the neighboring groups, probably not in the same detail. It is especially fortunate that the Local Group contains a large number of elliptical galaxies, from the smallest dwarfs to at least intermediate E types.

The study of the nearby galaxies shows very strikingly the intimate relation between stellar populations and dust and gas. This relation, of course, leads immediately to the problem of stellar evolution. Certainly we have taken only the very first steps in this direction; much work will be necessary, on both the theoretical and the observational sides, before we reach solid ground. I need only mention such problems as the evolutionary time scale, which is still wide open at present.

But now we can at least see the evolutionary picture of a galaxy in rough outline as far as the stars are concerned. At least roughly (let us not claim more) we can arrange certain groups in temporal order without paying too much attention to what the actual time differences are. It seems to me that we have excellent evidence today that star formation started at about the same time in all the galaxies of the Local Group. If we put it the other way around, the oldest stars that we can find in all these galaxies are the halo stars, such as the cluster-type variables. Even in the Magellanic Clouds we have convincing evidence of the presence of cluster-type variables, and we know that systems like IC 1613 and M 33 contain fast novae. Let me also say that Sandage has shown me pictures of M 33 taken in the infrared, on which one can see most beautifully the emergence of Population II stars between the spiral arms. So there is excellent evidence that star formation started at the same time in all these systems. There are elliptical systems, from dwarf to intermediate, Sc systems, dwarfs and irregular galaxies, and, according to the evidence we have now, stars in all these systems are of about the same age; that is a remarkable thing.

The second point I should like to make is that the bulk of the stars in all these systems are old stars, except perhaps in the Magellanic Clouds. In the elliptical galaxies and in the spirals — I would even go so far as the *Sc*'s — the bulk of the stars are old. This is true of their numbers, and even more true of their mass. We know that star formation has practically stopped in the *E* galaxies: there is too little dust and gas present, and we see no supergiant stars except in such pathological cases as those in the clouds near the center of NGC 205. This, on the other hand, is one of the nicest proofs that the moment dust and gas are present star formation is practically immediate.

Since all the galaxies of which I have just been speaking are members of the Local Group, we must conclude that they are of common origin. We must treat groups of galaxies in just the same way as star clusters and star groups in our own Galaxy. Such groups may have lost members in the course of time, but there has been practically no chance for them to acquire new members. So the galaxies of the Local Group must have been members from the beginning, which means that they very probably had a common origin as galaxies. Thus we know two things about the Local Group: as stellar systems, all these galaxies, whether they are irregulars, spirals, or ellipticals, are most likely of the same age; and star formation must have started at the same time in all of them, because in all of them we can identify old Population II. This Population II may, of course, vary in strength, and one of our big problems is to find out the relative strength of Population II and Population I in the Magellanic Clouds. On this important problem we have very little evidence at the present time. The fact that Thackeray and Wesselink have found quite a number of cluster-type variables outside NGC 121, which are clearly not members of the cluster itself, indicates that we are going far enough down in magnitude, and that this old population is showing up in a way that can be handled. So when we know the total number of cluster-type variables in the Magellanic Clouds, we shall have made the first step in this direction.

I am very certain that it would not be too difficult to observe the sheet of Population II stars if one studied the proper region on the outskirts of the Magellanic Clouds. Nobody in his senses would attempt it in the central bar, where the star density is too overwhelming, but should search between the bar and the outskirts. If I had to do it, I should choose a region around one of the globular clusters, selected by means of its color index. We know the upper limit of brightness for which to look, and I should simply see whether the sheet of stars that shows up on red plates shows the same magnitude as the brightest stars in the globular clusters. They will really stand out on photovisual plates, because their absolute magnitude is -3^M. They have not been found on red-sensitive plates because these red plates are not especially fast.

I have made two points: all the members of the Local Group have been members from the beginning, and probably have a common origin; and star formation started at about the same time in all of them, since the oldest stars that we find are always metal-poor stars of what we call the halo population in our own Galaxy. A third point is that the bulk of the stars that exist today in these galaxies were formed at an early time, and are old stars. This is certainly true for the E galaxies and the Sb spirals. We know that star formation is over in the E galaxies today. This suggests that the original rate of star formation depended on the nebular type, and the nebular type depends essentially on the state of rotation. The rate of formation may have been slow for the Magellanic Clouds, very fast for the other systems. The question of how much of the mass of the Clouds is in the form of gas is one that needs to be solved. In both Clouds the gas lies on the outskirts, and some radio observers fear that the present determination of the mass of the H I regions in the Clouds is too high because it rests on integration over a large area. The observational values are affected by considerable uncertainty.

We have, then, the picture that star formation is over in the *E* galaxies. If we stick to the *Sb* and *Sc* systems, where we have observations, we can say that the bulk of the stars was probably formed at a very early epoch. As to how much gas is still available in these galaxies, the data from the Leiden observations, at least for H I, indicate that only 1 percent, or at best 2 percent, of the matter is still available in the form of gas for star formation. This would mean, as far as these systems are concerned, the end of star formation.

We have of course no observations of this sort for galaxies at large. We must now rely on the fact that I mentioned at the beginning of this chapter, that, wherever we look in space, we find galaxies of the same sort as those represented in the Local System. We cannot get around this fact. We have now to rely entirely on the argument that there is an intimate relation between form of galaxy and its content of gas and stars. Since, in spite of all our searches, we have found no systems that do not fall within our classification of galaxies, we must assume that the Local Group is representative of galaxies at large. If we accept this, I think we can come to an evolutionary picture of the universe.

We have to ask ourselves the question: is it reasonable that star formation started in all galaxies at about the same time? Let me say that my preference, speaking as an observer, is the following. The facts indicate that before this epoch of star formation there was another epoch that determined the nearly simultaneous appearance, or formation, of stars in galaxies. Let me illustrate by saying what kind of system I was really interested in, and looked for. Take the example of an E galaxy in a very much earlier state of development, when it contained much gas. According to the meager data that we have today, we should expect that at an earlier date, when gas was still present and star formation was going on at a great rate, the galaxy would have been loaded with supergiants and H II regions. Such a galaxy, even if it were at a considerable distance, would stand out right away on account of its speckled appearance. It would be smooth, except for the H II regions, which would stand out especially on filter plates, which show up the emission lines. Although I was very keen on finding such systems, and hunted for them, I found nothing of the sort.

Simply from the observational side, then, I arrive at the following picture. We are situated in the Local Group, which has many members, and these members are kept together by gravitation, just as in a star cluster. As galaxies they must be pretty old, and they show a whole array of types, from E to Sb and Sc. It is really surprising to me that, within certain limits, star formation set in apparently at about the same time in all of them. I believe that this simply points to an event which preceded the star formation and determined it. I should rather like to conclude, since the galaxies in the Local Group are all equally old, irrespective of their type, and since the time scale that we use at present cannot be increased by several hundred percent, that there is very little room for evolution of the *forms* of galaxies.

Let us turn to the M 101 system, which we can study

quite decently. It contains many galaxies, including two Sculptor-type systems and one of Magellanic type. But most of the dwarfs are small galaxies of very regular outline. The larger ones show spiral structure; the arms are rather thick, but if you look at them you will say that they show spiral structure. But there is a series of smaller and smaller ones, and suddenly there comes a point at which, if I am honest, I should not say there was spiral structure. Finally there is definitely no spiral structure. Here you have small systems, apparently old, because they are quite regular. Their brightest stars belong to Population I, but are not very luminous. One certainly gets the feeling that spiral structure becomes inconspicuous for systems below a certain size, even if gas is present, simply because the amount of distortion by differential rotation is too small. You have a series of spirals of decreasing size; at a certain point you see, at best, 20 percent of an arm, and in the smaller ones you can still recognize a nucleus, but nothing more. In the M 101 group these systems with small disks continue down to a very small size.

We have come to the point where the evolution of galaxies themselves has become important. It would be highly desirable that the development of a gaseous galaxy should be followed up theoretically, at least in a crude way. So far as I know, very little has been done on this problem; we have the old investigations about liquid equilibrium figures, but I think they are not of much use in this problem. With our observations we have reached a point where we are simply unable to draw any definite conclusion, unless the theory helps us. I hope some day there will be action, because otherwise we are lost.

CHAPTER 20

THE STRUCTURE OF THE GALAXY

We now turn to our own Galaxy, to see how far we can understand it on the basis of what we have learned from other galaxies. For convenience we can divide it roughly into the disk, the surrounding halo, and the central nucleus.

Let us begin with the galactic halo. When Shapley had completed his study of globular clusters, his picture of our Galaxy consisted of the disk and a system of globular clusters that are concentrated toward the nucleus. In one way it was a curious picture — a disk of stars surrounded by a halo of clusters. An attempt to find whether there were other constituents of the halo resulted in the discovery that isolated cluster-type variables were also members of it. In the early 1920's I was studying globular clusters with a very large field. The first one was M 53, which contained about 50 cluster-type variables, with very well-defined mean brightness. But suddenly a number of other cluster-type variables showed up in this high-latitude field, some of them 3 magnitudes brighter than those in the cluster, some 2.5 magnitudes fainter. When I studied other clusters I obtained the same sort of result. So clearly the halo contained individual cluster-type variables as well as the clusters. At the time there was an idea that these isolated cluster-type variables were escapees from globular clusters, but this argument could easily be refuted, because the number of variables was so large that a huge number of globular clusters would have had to disintegrate to provide them. Shapley and his collaborators have investigated the galactic halo in this way in many directions, and have confirmed that cluster-type variables are present in every field.

The results of these Harvard investigations have been discussed by Shapley himself, and also by Kukarkin. Probably the most elaborate discussion is that of Kukarkin, and I will give a few of his results, and show that they do not

Table 18. Number of cluster-type variables per square degree in the direction of the galactic pole.

Magnitude (m)	$A(m)_{90°}$
10	0.00116
11	.00286
12	.00635
13	.01617
14	.03064
15	.03568
16	.00171

tell the full story. Table 18 gives the number of cluster-type variables per square degree in a 1-magnitude interval in the direction of the galactic pole, $A(m)_{90°}$. If log A (m) is plotted against m, the curve is seen to fall sharply near m = 16. Kukarkin apparently thinks that maximum numbers are reached for m a little fainter than 16, but clearly the last value is very incomplete, as the limiting magnitude with the telescope used is about $16^m.5$ to $16^m.7$.

From this material, Kukarkin derived the number of cluster-type variables per unit volume, log N, in the form:

$$\log N = 7.289 - 0.000219|\bar{z}|,$$

where z is the distance from the galactic plane in parsecs, and the unit of volume is the volume occupied by the cluster-type variables between $9^m.5$ and $10^m.5$. But the solution is not quite so simple as he assumed; there are strong indications that we really have a mixture of two groups of stars, one with rather a strong concentration toward the galactic plane ($|z| < 5$ kpc) and another for stars more distant from the plane. The picture is not as simple as Kukarkin assumed, and there are not enough stars for a satisfactory solution.

A new program has been undertaken at Groningen for studying the distribution of the cluster-type stars. One thing we can certainly assume about the galactic halo is that it must possess very high rotational symmetry, so that we can restrict the investigation to the meridian plane, perpendicular to the plane of the Galaxy, that contains the sun and the galactic center. That is important, because any investigation of this sort will be very time-consuming, and we must restrict ourselves to a minimum number of plates. The fields were chosen so that, if they contain absorption, the absorption is at least as homogeneous as possible.

The plates are to be those taken on Palomar with the 48-inch Schmidt with a field $6°.6 \times 6°.6$, and 25 to 30 pairs of plates are to be intercompared. To give an example: from the first four plate comparisons, 400 variables were found on each pair, 1200 variables from four comparisons. Variables are found in large numbers down to 20^m; with allowance for absorption, this means a limit of 17^M.5, corresponding to a distance of 32 kpc.

The program is a very ambitious one, and it will have to continue until no new variables are discovered by successive plate comparisons. I think that it will take care of the density distribution in the galactic nucleus, as far as cluster-type variables are concerned. But the Groningen staff will have their work cut out for them if they want to complete it in 7 to 10 years.

We should also like to know what is the real make-up of the halo, and here we have practically no data. How many long-period and irregular variables are there? Another program has been planned to cover these, and also to determine fundamental photoelectric sequences for all the fields, so that the results should be on a well-defined photometric system. Later it will be necessary to obtain the color excesses of the cluster-type variables in low latitude, and this again will be a major problem.

There are other programs, such as the one in which Haro and Luyten are picking up the blue halo stars of the horizontal branch and its extension. These blue stars of the horizontal branch run down to the white dwarfs, so they represent absolute magnitudes from 0^M to +9^M or even fainter. These stars will not contribute much to investigations of structure, but they will throw interesting light on the make-up of the halo. Haro is continuing a third program: the search for planetary nebulae outside the plane of the Galaxy on objective-prism plates. He has found some very faint planetary nebulae in the galactic halo, at high galactic latitudes and large distances, with diameters of 5″. This program also will be very important in the whole physical picture. We know very little about the halo yet, and there is plenty of work to be done.

Let us turn now to the region of the galactic center. In one way we are in a very much better position than we were 20 years ago, because radio astronomers have furnished valuable data on the location of the galactic center. Table 19, which gives the best data available today, shows that we have determined the position of the galactic center region with some accuracy.

Let us start with the optical data, which are still good. Here one can only determine the longitude with sufficient accuracy. I give one of the more recent figures for the planetaries, which are strongly concentrated to the galactic center; Minkowski has almost tripled the known number. For the radio data I have taken only those that refer not to the 21-cm wavelength, but to a different part of the continuum, determining the longitude of the point of maximum intensity in the plane of the galaxy. The radio observers are in a favored position because the absorbing clouds play only a very mild role. I do not give the details of the four determinations; the beam widths range from 25° to 0°.8, and curiously enough it does not seem to make a difference whether one uses a wide beam or a narrower one; the narrower ones show more of the detailed distribution, but the determination of the intensity maximum is just as difficult, just because of the detail. Recently Gum and Pawsey, who made a new determination of the location of the plane of the Galaxy from the distribution of the hydrogen, combined all these optical and radio determinations and gave the value 327°.5 ± 1°, which is our best value for the longitude of the galactic center at the present time.

Table 19. Position of the galactic center.

Observer	l	b	Frequency (Mc/sec)	Beam width
Optical				
Globular clusters (Shapley)	$325° \pm 3°$	—		
Planetary nebulae (Minkowski)	$328° \pm 3°$	—		
Radio				
Four investigations	$328°.0 \pm 1°$	—		
Gum and Pawsey	$327°.5 \pm 1°$	—		
Sgr A				
Kerr and Hindman	$325°$	$-1°.4$	1400	1°.5
Hagen and Lilley	$327°.8$	$-1°.4$	1400	0°.9
Mills	$327°.9 \pm 0°.2$	$-1°.5 \pm 0°.2$	85.5	0°.8
Westerhout	$327°.71 \pm 0°.03$	$-1°.47 \pm 0°.03$	1400	0°.6
Haddock and Mayer	$327°.6 \pm 0°.1$	$-1°.5 \pm 0°.1$	3200	0°.4
Mean	$327°.8$	$-1°.45$		

So we know that the value for the general intensity distribution is like that of the globular clusters and the planetary nebulae to within something of the order of 1°.

Now we come to the specific radio source, Sagittarius A, which we have very good reason to believe is actually the galactic center. Now we can get both latitude and longitude. All the figures in Table 19 are referred to the old Lund system, Ohlssen's pole. Here we have entirely different accuracy: Mills gives a probable error of 0°.2, and Westerhout gives it to hundredths of a degree, and these accuracies are not fictitious.

The value −1°.45 for the galactic latitude of Sagittarius A puts it not only in the same longitude as the galactic plane, but exactly on the galactic plane, since the correction from the Ohlsson system to the new galactic plane is exactly −1°.45.

There is good reason for identifying Sagittarius A with the galactic center. The arms closest to the center that have so far been found at Leiden are two arms that show an outward velocity of about 50 km/sec. Counting from our position, this would be the third inner arm. There is a corresponding arm outside, with positive velocity of the same order. Now the first inner arm absorbs the 21-cm Sagittarius A source, whereas the other shows no absorption. Since the Leiden people believe, on account of the velocities, that the first arm must be less than 2.5 kpc from the galactic center, the conclusion that Sagittarius A is at the galactic center is unavoidable.

We may believe that this is a case similar to what we see in the galaxies that Seyfert investigated, Sb's with strong wide emissions at the centers. And actually some of the nearer ones are radio sources, although we are not quite sure that all the radio energy is coming from the nucleus.

Now that the position of the galactic center is known so accurately, it should be possible to locate it. The first thing I did was to photograph this very region at the 48-inch Schmidt with very long exposures in the red, using a filter to cut out Hα partially so that I could expose for 5 hours. I got a very beautiful 103a-F plate, which goes still further than 103a-E, of just the right density — not too much sky fog. I took a second exposure and then a third, and finally with the 48-inch Schmidt I took a very perfect plate with an exposure of 7 hours. Of course now I could locate exactly what came out. And what came out was a number of globular clusters, which cannot be photographed at all on the blue plates or photovisual plates, because they are so heavily reddened. A little bit farther north there is an emission nebulosity, quite well known, in one of the nearer arms. But this is not Sagittarius A; it is farther along the plane of the Galaxy toward higher longitude.

Not giving up yet, I decided to use the 200-inch, and covered the area, assuming the error in either coordinate to be ±0°.1. I covered an area three times as large with red and infrared plates. They are not all of the best quality, but I should grade the worst as seeing 2–3, and I have a few where the seeing was 4. And there is not the slightest indication of anything except a faint sheet of stars, and more globular clusters, exceedingly reddened, which could all be in front.

I had already attempted the same thing several years ago, and found a number of exceedingly reddened globular clusters. Since Whitford had always wanted a good globular cluster that was so heavily reddened that he could determine the run of the absorption with high accuracy, I gave him one of these clusters. It is very easily identified because it lies just midway between two eighth-magnitude stars; you cannot miss it. I should say that if there were no absorption the cluster would extend beyond both the stars; it is just between them. So Whitford tried it; he could get no response in the ultraviolet (which was not surprising); he could get no response in the blue; he could get no response in the yellow. He got one only in the red. So the attempt to determine the color went awry. But we sat down and figured out that the absorption for this cluster must be at least 7 magnitudes. I have not the slightest doubt now that the absorption in front of the nucleus itself must be of the order of 9 or 10 magnitudes, so I am satisfied that with our present means we are sunk. When image converters can really furnish another factor of 10, then it might be possible to do the thing with the 200-inch. There is one unexplored possibility, but I do not believe in it too much, and that is to go over the region carefully between 7000 and 7200 A, which is exceedingly weak in night-sky bands; perhaps something would appear in there. But it would be just throwing in pennies when you need a hundred dollars.

In former years I was searching for something like a nucleus, a sort of superglobular cluster. I always counted on huge star density, increasing toward the center, rising like a steep hill, so that it would be observable if one could get really good seeing. But now I have given up hope that any peak would help, even if it ran up sharply like a needle. I looked for it, but the only things that you see are bonafide globular clusters; those you see right away.

You have only to look at the nucleus of the Andromeda Nebula, say on a plate, taken with the 100-inch or the 200-inch, and you see that it is surrounded by a large number of globular clusters which are concentrated there. Take one of the densest globular clusters and compare it with the dense image of the nucleus, and you will see that it is of an entirely different order. A nucleus of this kind, if it could be observed for the Galaxy, would stand out because it would not be resolved — the stars would be so jammed up. Once I thought I had it; looking at the plates I thought that one of these objects could be it. But when I saw the stars still so well separated I realized that it was one of these globular clusters that we still see in the foreground, heavily affected by absorption.

From the work of Pawsey and Gum, determining the plane of the Galaxy, we know that in the inner parts of the Galaxy up to 8 kpc the arms lie really marvelously close to the plane, and so does the dust. Three arms lie one behind another, you look at an angle of 0°, and you have had it. If you look at a Schmidt plate through which the galactic equator runs, especially when you compare a red plate with a blue one, you see a band along the galactic equator in which there is no structure whatsoever. As you go outside the band you can see that different layers are concerned, but there is nothing in this inner band, which has, I would say, a width of 2°; it is just a uniform blocked-out area. I probably spent more time on the problem than I should have done, but I know one thing: I cannot do it. But I have not the slightest doubt that Sagittarius A is the galactic nucleus.

Since we cannot get at the nucleus itself, there is nothing for it but to study the bulge, and to try to get as close in as you can. You know that a number of attempts have been made, but I think there has been a confusion of terminology. Stebbins and Whitford claimed that they had got the galactic center; so did Kalinyak, Krassovsky, and Nikonov, and so, more recently, did Dufay. One should probably say the region around the galactic center, the bright region that sticks out beyond the nebulosities, for it is no more than that. One can get these same regions on infrared film with a Leica camera attached to the telescope, and what they are is a matter of terminology.

That the galactic bulge is full of variable stars was, I think, first found at Harvard and at Leiden, and in the 1930's it was literally plastered with variable star fields. However, the work was done with small telescopes with limiting magnitude about 16^m.5. With Shapley's value of 10 kpc for the distance to the galactic center (the best available at the time), the distance modulus would be 15^m. All these programs were started with the idea that there was no absorption whatever, but now we know that the absorption is very respectable, and at low latitudes, even in the brightest part of the Sagittarius Cloud, it is of the order of 3 magnitudes. So most of this work, though some of the fields turned out to be very rich in variable stars, did not contribute to our knowledge of the structure of the nucleus, or to the distance of the galactic center — with one exception that I shall mention later.

When I got interested in the galactic bulge in 1938, I fortunately had the 18-inch Schmidt on Palomar at my disposal. At that time I photographed the whole galactic bulge on blue and red plates from longitude 320° to 350°, and within 20° on either side. I went over the plates as closely as possible in order to pick out the best regions, and selected three: 329°.1, −4°.0; 331°.7, −6°.6; and 328°, −4°.3. Afterward I went over the plates again independently, and selected the same regions again; I am certain that they are the best regions if you want to study low latitudes. These galactic latitudes and longitudes are on the system of van Tulder, which is very close to the new system. The first two fields are very homogeneous, especially the second, which is the best of all. *De facto,* I selected the third for my own investigation.

The region I actually chose does not compare favorably with the other two. My reason for selecting it was that it was clear that the most important problem would be the determination of the absorption. In the very homogeneous fields I should have had to rely on determining the color excesses of the cluster-type variables, with all the difficulties involved, in a field that is very far south for Mount Wilson. On the other hand, NGC 6522 is very likely embedded in the nuclear region itself. Stebbins and Whitford had already determined a color excess for this globular cluster, and moreover were willing to make more extensive measures for me, and I hung on to this one globular cluster for dear life, even taking into account that the whole field was not as homogeneous as I could have wished.

In the next chapter I shall discuss this field, and show that the surmise that NGC 6522 is embedded in the galactic bulge was correct, that (with certain uncertainties) it furnishes the absorption, and that this absorption can be eliminated in a very straightforward way. But whether the problem has been completely solved is another thing.

CHAPTER 21

THE GALACTIC NUCLEUS

As I mentioned in the last chapter, I preferred to pick the field centered on NGC 6522 for my investigation of the galactic center because the blue and red plates that I had already taken with the 18-inch Schmidt on Palomar showed strong reddening all over the Sagittarius Cloud. This meant that we must deal with strong absorption, and in this case the absorption could only be determined by way of NGC 6522, for which Stebbins and Whitford had made a preliminary determination of color excess. They afterward repeated the determination with four colors, and derived the large photographic absorption of $2^m.75$. It seemed quite safe to derive the absorption from the color excess of the cluster, since the appearance of the plates — especially those taken with the 100-inch — showed that NGC 6522 was obviously embedded in the nucleus itself.

In previous years, observers with smaller instruments had plastered the galactic nucleus with plates, and had found it to be filled with cluster-type variables. Unfortunately they could not find the magnitude at which the cluster-type variables have maximum frequency, because that magnitude was so close to the limit of their plates. Therefore I was very careful that the same thing should not happen in my work. Since I knew from the color of NGC 6522 that I had to count on about 3 magnitudes' absorption, and, considering that the distance of the galactic nucleus might be anything from 8 kpc to 12 kpc, the frequency maximum must lie between about 17^m and 19^m. So I decided to use the 100-inch telescope, with which I could easily reach a limiting magnitude of $20^m.5$ in 25 to 30 minutes, so that the limiting magnitude would not affect the frequency maximum.

I observed the field from 1945 to 1949 with the 100-inch, and in those years I obtained 137 plates. This could be regarded, of course, only as a first attempt, because the region is at declination $-30°$, which means a zenith distance of $65°$ from Mount Wilson on the meridian, and this limited the hour angle severely. I could not use hour angles larger than $2^h.5$ east or west. All the images are slightly elongated, and even more so as you go away from the meridian. Fortunately we usually have very good seeing conditions on Mount Wilson during the months of July, August, and September, when the field is most easily accessible, and the plates are really of excellent quality when the large zenith distance is taken into account.

A rather thorough search for variables was made on 27 pairs of plates, all very well matched, with the blink comparator, and 285 variables were finally found. The plate was 5×7 inches, and the field was $42'.0 \times 31'.8$, with an area of 0.3715 deg^2.

A sequence was selected for each variable, and the next

step was to bring all the sequences into a homogeneous system. The mirror of the 100-inch was diaphragmed down to 58 inches, which gives a coma-free field of 30′ diameter; four overlapping fields and a central field were centered on the mean of the diagonal axis of the area to be studied. Everything was finally referred to four primary sequences, each running from about 14^m to $20^m.5$, and the four sequences were connected with Selected Area 68 by a large number of transfers. For the magnitudes in Selected Area 68 I already had the photoelectric magnitudes of Stebbins and Whitford down to $19^m.0$, and shortly afterward Baum's measures down to $21^m.5$ became available. From the transfers of Selected Area 68, the mean error of the zero point of the magnitudes at the present time is of the order of $0^m.05$. To this accuracy we are on the International System.

Before going any further we must decide how far the field covered by these 5×7-inch plates is homogeneous. I have already mentioned that it is not as good as the two other Sagittarius fields, especially because there is already absorption in its outer regions. After the variables had been estimated and studied I divided it into a number of subregions. Counting from the center, where NGC 6522 is situated, the central region had a radius of $7′.44$, and the next ring of equal area ran from $7′.44$ to $11′.23$. In the inner ring I found 44 variables and in the outer ring, 47; the numbers are about equal. If we proceed in bigger steps, we find 91 variables between $0′$ and $11′.23$, 91 variables between $11′.23$ and $15′.88$, and 70 variables beyond $15′.88$. These three regions are of the same area, and there is no loss in the number of variable stars in going from the first to the second; the third area shows the effect of the absorptions. I should add that the six variables in NGC 6522 were excluded from these counts, since they are not field variables. We conclude that if we stay within 15'.88 of NGC 6522 there is no loss of variables due to absorption or coma, and that we can use the variables within this area to study the statistics of the various types.

The estimation of the variables and the determination of the light curves was carried out by Sergei Gaposchkin. Table 20 summarizes the distribution of his material

Table 20.   Distribution of variable stars in the galactic nucleus.

Type	Percent
RR Lyrae	40.7
RV Tauri	2.7
Long-period variables, $P < 220^{\mathrm{d}}$	9.9
Mira variables	6.0
Semiregular and irregular	22.5
Eclipsing	13.2
Miscellaneous and undetermined	5.0

according to type. The classification scheme is that of Kukarkin, Mira variables being the long-period variables with amplitude larger than about $2^{\mathrm{m}}.5$.

We see at once that the cluster-type variables are by far the dominating type. For the long-period variables I should prefer not to make this artificial separation, but to wait until we have spectra, which should not be too difficult to obtain with an objective prism on one of the Schmidts. The groups overlap, and cannot be distinguished by photometry alone; we need to know whether the stars have the typical emission spectrum of the Mira stars. Perhaps a small percentage of the stars with period less than 220^{d} will prove to be Mira stars, but it will probably not be larger than 10 or 20 percent. The semiregular group is always ill-defined; some of these stars may be long-period variables, some may be RV Tauri stars, and one can clean them up if spectra are obtained. Many stars with periods of 80^d up to 150^d or 200^d, with amplitudes of $1^m.0$, $1^m.5$, sometimes $2^m.0$, are found in globular clusters; they are quite regular, but though they show emission lines they do not in general have the typical Mira spectrum.

The high percentage, 13.2, of eclipsing stars is very interesting. When we discussed globular clusters I mentioned that they undoubtedly contain eclipsing stars. If we adopt absolute magnitude $0^M.0$ for the cluster-type stars, as we shall do in what follows, the group of eclipsing stars in our field has an average absolute magnitude $+0^M.5$.

The dominant group are the RR Lyrae stars, and they are very remarkable. The frequency distribution of their periods is definitely different from what we find in our Galaxy as a whole, or in fields that have been studied by other observers in higher latitudes. For example, if we compare our field, at galactic latitude $-4°.3$, with van Gent's field at galactic latitude $-18°.5$, we see that in van Gent's field the mean period is close to $0^d.5$ and the number falls off steeply for periods less than $0^d.45$, whereas in the field around NGC 6522 the mean period is around $0^d.33$. Normally the light curves for periods around a third of a day are sine curves of small amplitude, whereas our variables have asymmetric curves with nearly normal amplitudes. So there are two differences between the nucleus and the other fields: in the nucleus the mean periods are shorter, and the light curves are asymmetric with nearly normal amplitudes.

On account of the limitations in hour angle, there will

always be uncertainties in period determination. Observations taken in the Southern Hemisphere, where one can study the stars for 12 hours at a stretch, would settle the doubtful cases, but, whatever the corrections may prove to be, I am confident that they are small.

We can now discuss the density distribution of the cluster-type variables around the galactic nucleus, and the distance of the nucleus itself. It is clear that we have no loss of variables within 15'.88 of NGC 6522, but this does not settle the question whether the absorption in the field is homogeneous. In order to test this point, I subdivided the material within 15'.88 of the center of the field into three areas, which contained about the same number of cluster-type variables; the areas are not quite identical. In each of these areas the magnitudes of the cluster-type variables show a well-defined frequency distribution. In all three, the stars begin at about 16^{m}.8, and the number drops to zero at about 18^{m}.1. The mean magnitudes of the cluster-type variables in the three areas are given in Table 21. Within the mean errors, Areas I and II show no indica-

Table 21.   Average magnitude
of cluster-type variables by areas.

Area	Limits	$\overline{m}$
I	0'.00– 8'.75	17.34 ± 0.07
II	8'.75–12'.38	17.40 ± 0.05
III	12'.38–15'.9	17.57 ± 0.06

tion of differential absorption, but Area III has definitely higher obscuration than the other two. So in further discussion of the cluster-type variables I now omit Area III, and restrict myself to a distance less than 12'.38 from the center of the field. This material should suffice for determining the density distribution and the distance to the galactic center.

The procedure is very transparent, and requires no further assumptions. The figures are given in Table 22. The first column gives our scale of apparent magnitudes, to which we apply the correction of $2^m.75$ for absorption (second column). If we assume a photographic absolute magnitude $0^M.0$ for the cluster-type variables, we obtain the distance given in the third column. We are shooting through the nucleus in a cone with angle 12'.38, and the fourth column gives the linear radius of the cone, for the corresponding distances. The volume of the cone defined by these distances is given in the fifth column, and the reciprocal of the volume in the sixth. The numbers of variables in these sections are given in the seventh column; with the aid of the sixth column you obtain the number of variables per million cubic parsecs (eighth column). Finally, the ninth column gives the mean distances at which these different cells lie, and the tenth shows the distances of these cells from the galactic center.

The highest density per unit volume is 6.93 per million cubic parsecs, which is reached at a distance of 8.14 kpc in the line of sight, and this volume is 0.6 kpc from the galactic center. Now we can immediately write down the distance of NGC 6522, which contains six cluster-type variables. The observed value of $m - M$ is $17^m.78 \pm 0^m.08$; corrected for absorption, $m - M = 15^m.03$, corresponding to a distance of 10.14 kpc. Thus the cluster is even farther away than the galactic center.

We have determined the distance of the region of great-

Table 22. The field of NGC 6522.

Magnitude		Distance (pc)	Cone radius (pc)	Volume (10^6 pc³)	Volume⁻¹ (10^{-6} pc⁻³)	Number of variables	Number of variables per 10^6 pc³	Mean dist. (kpc)	Dist. from galactic center (kpc)
Appar. (m)	Corr. (m)								
16.80	14.05	6457	23.2	1.164	0.859	1	0.86	6.768	1.501
17.00	14.25	7079	25.5	1.523	.652	6	3.91	7.421	0.945
17.20	14.45	7763	27.9	2.021	.495	14	6.93	8.137	.616
17.40	14.65	8511	30.6	2.666	.375	9	3.38	8.922	.998
17.60	14.85	9333	33.1	3.513	.285	5	1.43	9.783	1.757
17.80	15.05	10233	36.8	4.631	.216	1	0.22	10.727	2.662
18.00	15.25	11220	40.4						

est density, which is of course the region closest to the nucleus. If its distance is D_1, and if D_0 is the distance to the galactic center, we find that

$$D_0 = D_1 \sec b \sec (l - l_0),$$

where b is the galactic latitude, l the galactic longitude of the field, and l_0 that of the galactic center, which can be taken from the radio observations. We then obtain for the distance of the galactic center:

$$D_0 = 8.16 \text{ kpc.}$$

We see from our frequency distribution (Table 22) that we are already far past the maximum at 10.7 kpc; since our globular cluster is already on the other side of the center, the absorption determined for the globular cluster should have fully taken care of the absorption for the central region.

We might indeed have overcompensated for the absorption, but when you consider the distribution of the cluster-type variables you can see that there is very little danger of overcompensation. If we take the inner region, we find that the faintest cluster-type variable has apparent median magnitude $19^m.34$; if we apply again the absorption correction $2^m.75$ (by this time we are far above the galactic plane), we find a corrected magnitude $16^m.59$, corresponding to a distance of 20.8 kpc. We subtract 8.2 kpc to get the distance beyond the center, and find that this star is 12.6 kpc beyond the galactic center on the far side. There is a similar star, a little brighter; these are the two most distant stars that have shown up in the field. This distance agrees exceedingly well with what we should expect from the most distant spiral arms; probably the outermost parts of the disk have a radius of 12 or 13 kpc.

The same thing follows from another argument. In the direction of the galactic rotation, the highest velocities that we find are of the order of 60 km/sec. From this again, as Trumpler and Weaver have shown, we conclude that the extent of our Galaxy in the direction away from the center is 4 or 5 kpc, which would lead again to the same dimensions. This agreement shows that there is nothing seriously wrong with the present discussion; if we had made wrong assumptions about the absorption there would have been real discrepancies.

The field I have been discussing lies in the direction of the Large Sagittarius Cloud, where the absorption has been studied by Bok and his collaborators. In the direction of the Large Sagittarius Cloud they found $1^m.3$ absorption at a distance of 1230 pc, and about $2^m.0$ at about 2150 pc, which, if these data are correct, means that two-thirds of all the absorption is already accounted for in the first third of the distance to the cloud. I believe that they are essentially correct, which means that most of the absorption is in our neighborhood. I think this must be so, for the following reason. When you observe the Sagittarius cloud on blue- or red-sensitive plates you see both absorption streaks and clear areas, but the clear areas are already highly reddened. You would never see this homogeneity if the absorbing material were not very near, and covering the whole cloud.

The cluster NGC 6522 is a most unusual one. I was very much relieved when Morgan showed a year or two ago that it is a halo-type cluster, with the characteristic feature that the spectral type estimated from the metallic lines is about three-quarters of a class earlier than that estimated from the hydrogen lines. This is very important, for if it had been a disk cluster, about which "we don't know nothing," we should not have known the intrinsic color, and so far we are well off.

But now we find something very remarkable. The cluster contains six cluster-type stars, and the most distant of these are usually a good measure of the extent of a cluster. The one farthest from the center is 101″ away, and I searched in widening circles to see whether I could find any other members; but I could not. On the basis of this star, the linear diameter of the cluster is only 10 pc. It has total apparent magnitude $10^m.40$, so the integrated absolute magnitude is $-7^M.4$; the normal diameter for a cluster of this luminosity should be 70 pc. The corrected distance modulus, $15^m.03$, is just a little bit more than that of the Hercules cluster M 13, so I took short-exposure plates of M 13 with the 100-inch for comparison. The difference is striking: NGC 6522 is so highly condensed that even with the best seeing conditions everything is jammed up and you cannot see through it, whereas with these short exposures the stars of M 13 are so far apart that the cluster seems full of holes.

In NGC 6522, then, we have a cluster of exceedingly high density, and I think that is inevitable because it is a halo cluster. Two kinds of object are concentrated at the galactic center — members of the disk, such as the disk clusters, and members of the halo, such as the halo clusters. Since there is a concentration, we know that a large number of these clusters must be moving in orbits with small major axes; otherwise, if they are moving in elliptical orbits, there would be no concentration. So actually the frequency distribution of the major axes of the globular clusters of the halo must have a very high peak for small major axes, which means that any halo cluster that we observe in the region of the galactic center has most likely always stayed in that region. And in order to survive it has to have a terrific density. Obviously NGC 6522 has a terrific density because it has such a small diameter, and the reason for the small diameter must be that it has repeatedly lost members because it moves in such a dense field. This loss of members (and of kinetic energy) has caused it to shrink and has made it increasingly stable against the surrounding dense field. I think that in NGC 6522 we have a most interesting case — a cluster of normal brightness that has a diameter one-seventh of what one would expect, that is apparently a permanent member of the central region of our Galaxy. You must imagine the central region to be like a huge supercluster, and this object is moving around permanently in the region; it really deserves more attention.

To summarize again: if the distance to the galactic center is 8.16 kpc, the true distance modulus $m - M = 14^{m}.56$. The uncertain feature of the determination is the fact that we had to rely on the color excess of NGC 6522 to get the total absorption. I think that, as the data stand today, a value of $2^{m}.75$ is unavoidable; you may add about $0^{m}.10$ or $0^{m}.15$, but that is all.

As a check, I have carried through a similar computation for Harvard Milky Way Field 269, where Shapley made an early investigation of the variable stars in order to determine the distance to the galactic center. He used a special series of Bruce plates that avoided the difficulty imposed by limiting magnitude. The galactic latitude and longitude of MWF 269 are $-19°.0$ and $307°.1$ on the van Tulder pole. As the galactic latitude is higher than what I used in Sagittarius, the number of stars per cubic parsec is of course smaller, and the data have to be combined in larger steps,

0^m.5 instead of 0^m.2. At galactic latitude 19° the extragalactic nebulae come through in large numbers, and Shapley derived an absorption of 0^m.61 from nebular counts. This is exactly the value that one would expect if the general absorption over the field has the old value of 0^m.26 at the pole. I have taken the data just as Shapley gave them, and, using his value for the absorption, obtained the results given in Table 23, which corresponds to Table 22, but without the first and sixth columns of the latter.

Just as before we find a density maximum with 37 cluster-type variables per 10^9 cubic parsecs. If again we apply to the distance the correction sec b sec $(l - l_0)$, we find for the distance to the galactic center the value 7.68 kpc, corresponding to a true distance modulus $m - M = 14^m.43$. Here, then, is a field in very much higher latitude; the absorption is smaller and the maximum number of stars is at a magnitude between 14^m and 15^m. The scale has meanwhile been shown to be correct by a check at −55° in the Southern Hemisphere. We see that the distance moduli determined at Mount Wilson and Harvard agree within 0^m.13. The distribution of the magnitudes of the cluster-type stars has a sharp maximum for my field, which is in low galactic latitude, and a much wider maximum for Shapley's field, in higher latitude.

The last columns of Tables 22 and 23 gave the distances from the galactic center for the respective sections; one can now plot the density of variables against these distances. Shapley's points for large distances fall off from the curve, on account of incompleteness, but for smaller distances they fall on the same line as those for the field of NGC 6522, even though Shapley's stars had normal periods and light curves, while those in the NGC 6522 field were

Table 23. The field of MWF 269.

Corr. mag. (m)	Distance (kpc)	Cone radius (pc)	Volume (10^9 pc^3)	Number of variables	Number of variables per 10^9 pc^3	Mean dist. (kpc)	Dist. from galactic center (kpc)
12.89	3.784	167	0.1097	2	18	4.27	4.40
13.39	4.764	210	.2189	7	32	5.38	3.88
13.89	5.998	264	.4367	16	37	6.77	3.62
14.39	7.551	333	.8713	27	31	8.53	4.02
14.89	9.506	419	1.7386	16	9.2	10.74	5.37
15.39	11.967	527	3.4689	5	1.4	13.52	7.65
15.89	15.066	664					

abnormal. The straight line that shows the density distribution as a function of distance from the galactic center can be very nicely represented by

$$\log N = 4.20 - 0.690\, R,$$

where N is the number of cluster-type variables per 10^9 cubic parsecs and R is the distance from the galactic center in kiloparsecs. This shows that the central system of our Galaxy cannot be highly flattened, so far as the cluster-type variables are concerned; it must rather be circularly symmetric. Let us take this relation as a first approximation.

We can now integrate over all distances and over the whole sphere. The result shows that there should be 66 cluster-type variables in a sphere of 50 pc radius around the galactic center. If we were there we should observe 66 cluster-type variables brighter than apparent magnitude $3^m.5$. There should be 21,250 within 1 kpc of the center, 61,100 within 2 kpc, 84,800 within 3 kpc, and by integrating up to infinity we obtain 99,300 cluster-type variables.

Let me say at once that this is not the total number of cluster-type variables in our Galaxy, but only the total number associated with the central system. This is a very concentrated system, comprising, within 3 kpc of the center, 85 percent of the cluster-type variables and presumably of the Population II that goes with them.

Let me make a final remark about MWF 269. In this field the long-period variables were worked up at Harvard in addition to the cluster-type variables. The frequency maximum for the cluster-type variables was near 15^m, and that for the long-period variables at maximum was near 13^m. We know that the magnitude for the frequency maximum of the cluster-type variables is correct, and it is the fainter of the two, so we can rely on that for the long-period variables. Here for the first time we get a good determination of the difference in absolute magnitude between the two groups. The long-period variables, which in this case have a mean period of about 225^d, are 2^m brighter at maximum than the cluster-type variables. In my discussion of the globular clusters I derived a difference of $1^m.9$, so the data now check very neatly; our efforts to converge to definite values have led to consistent results and there are no longer any contradictions.

CHAPTER 22

THE GALACTIC DISK

I N H I S study of the Andromeda Nebula, Morgan found evidence on McDonald plates that the brightest stars seen there belong to Population II, metal-poor stars like those of the galactic halo. But the integrated spectrum was that of G and K giants with strong cyanogen bands. His argument — and I think it is a good one — rested on the strength of the cyanogen bands. In giants of Population II these bands are weak, so he concluded that most of the light in the Andromeda Nebula does not actually come from metal-poor Population II stars but from an old population in which the metals are enriched, rather normal stars like those of M 67.

It was therefore very interesting for him to try the same thing in the center of our own Galaxy, and he obtained a curious result. He states that the integrated spectrum comes from dwarf stars of solar type, which means again about type G. He states expressly that these dwarfs are not of the halo type, which means that they are not metal-poor. Morgan's result is queer for two reasons. First, most of the luminosity comes from dwarf stars. With any plausible luminosity function that rises to a maximum, it is well known that most of the light comes from the first three or four magnitudes, so that in our galaxy, in contrast to the Andromeda Nebula, this indicates a very queer situation. The second queer thing is that these are not halo-type stars. He must use the metal lines in some way for classification, but the statement that the stars are dwarfs must rest on the fact that the cyanogen bands are weak; this is the case for all dwarf stars, whether they are metal-rich or metal-poor. So his classification must mean that he finds the metallic lines relatively strong.

I do not know what to make of Morgan's result. He seems to have made the assumption that galactic rotation will not cause trouble with the radial velocities, and that is certainly true, but on the other hand the internal velocity distribution in this direction is very large, as can be seen from the planetary nebulae.

At the present time, then, the spectroscopic observations are very difficult to reconcile with the photoelectric observations, which were done with the greatest care. The photoelectric colors are compatible with the color of a population of either metal-poor or normal stars, for the integrated color is not very sensitive in this respect; they could correspond to halo stars, a halo cluster, or a cluster like M 67.

When we turn to the disk of the Galaxy we are really in trouble at the present time, because we know so very little about it. From the very best data available — the spiral structure measured at 21 cm — we can make a good guess that the disk has a diameter of about 25 kpc. We are very much better off than we were ten years ago, but the picture is still a first approximation because everything is based on the adopted rotation law. It is very desirable to determine the positions of the various spiral arms by different methods, such as the distances of O and B stars. Such investigations should really be regarded from another point of view: if we can grade the distance indicators according to age, and use small ages of a few hundred million years, we may be able to discover differences in different regions.

One of the first results of such work is a paper by Becker and Stock, which compiles recent work on a number of open clusters and attempts to give a picture of their distribution in the solar neighborhood. Here we can just begin to see the emergence of the Perseus arm and the next arm on the inside, in addition to the arm in which the sun is located. But this picture itself shows how many more data we need even for nearby regions. Probably the most suitable objects for this purpose will be the O and B stars.

It is especially necessary to determine the location of the sun in the spiral arm. In this respect the radio observations are quite indeterminate. All the data indicate that the sun lies rather close to the inner edge of the arm, but it is very desirable to find the location more exactly.

I think that the recent data have led to two exceedingly important results. One is the distribution of the gas perpendicular to the galactic plane. The Leiden data can be very beautifully represented by a Maxwellian distribution with a half-width of 110 pc perpendicular to the galactic plane. This result applies to the vicinity of the sun, possibly up to a distance of 1 or 2 kpc, and there is reason to believe that it can be extended to the whole of the inner part of the Galaxy; at least it includes the next two arms.

The second result refers to the distribution of the Cepheids, a group of very bright, young stars, and has now been obtained reliably for the first time. The Leiden observers have recently determined the color excesses of southern Cepheids down to 12^{m}.5, and from these data it is possible to derive the distribution, perpendicular to the Galaxy, of Cepheids essentially within 2.4 pc. Walraven, Muller, and Oosterhoff have shown that here again the distribution is essentially Maxwellian with a half-width of 65 pc. The stars are more concentrated to the plane of the Galaxy than the gas itself, and this is eminently reasonable; the stars are in regions of higher density.

Here, therefore, we have for the first time figures that give us an idea of the thickness of the spiral structure. It is quite obvious that the spiral structure inferred from the gas, or from the stars that have recently formed from the gas, is a very flat and thin affair. As time progresses we shall get more data; there should be no difficulty in doing the same thing for the B and O stars in the next few years.

Now concerning the rest of the disk (and leaving out for the present the complications produced by the nuclear region) we are especially interested in getting a rough picture of the ages and chemical composition of the stars that make it up. De facto, we know very little about this.

Let us first consider objects that can be seen throughout a large part of the disk on account of their brightness — at least the part of the disk that lies on our side of the galactic center. Objects of this sort are the globular clusters that are concentrated to the disk, those to which Morgan called attention because they differ from the globular clusters of the halo in having metal lines of normal intensity. We saw saw in Chapter 12 that these clusters are not spread throughout the halo, but concentrated in the disk. At the present time we know very little about them, only that they are distinguished by having metallic lines of normal strength.

The novae are a second group of objects that we can see over large distances. Their distribution has been investigated by McLaughlin and by Kukarkin. Investigations of the novae in the Andromeda Nebula by Arp have shown that there is a close relation between duration of maximum and luminosity, so absolute luminosities are not difficult to obtain for novae with good light curves. Correction for absorption is more difficult; McLaughlin and Kukarkin made it in a roundabout way, the best that can be done. Their distance scales differ somewhat, because Kukarkin makes larger corrections for absorption, but the general picture is the same. Practically all the novae are found to belong to the disk of our Galaxy; all lie within about ±1000 pc of the galactic plane, about the thickness that we should expect. We know that novae are metal-poor stars of Population II, and we are therefore quite certain that old stars of halo type are now present in the galactic disk. I am not here speaking of the recurrent novae.

A third group of objects that we can observe over very large distances in our Galaxy is the planetary nebulae, again Population II objects. If we plot the known radial velocities of planetary nebulae against the galactic longitude we find a high concentration at 327°. Not only is there a high concentration, but also the planetary nebulae show a huge velocity dispersion as soon as you get into the center of the Galaxy. In this case we are very fortunate because even in the presence of absorption we can tell whether an emission-line object is a Be star or a planetary. Just like the novae, they show a strong concentration toward the galactic center; in fact, the concentration does not stand out so strongly for the novae.

So we have data on three groups of objects in the galactic disk: disk globular clusters (enriched clusters that are as old, or nearly as old, as metal-poor globular clusters); novae of Population II; and planetary nebulae. All these groups consist of very old stars, but we have a mixture: some are metal-poor, some are enriched.

I now come to the most conclusive group of all — the cluster-type variables of the disk. I mention this especially because they are usually regarded as a spherical system. But the fact that the number of cluster-type stars in a given volume can be represented in the form $\log N = a - bz$, where z is the distance from the galactic plane, shows that they are concentrated to the galactic plane. Such a relation would of course not exist if we were dealing with a spheroidal system.

Table 24.   Period and velocity
of cluster-type variables.

Period (d)	Mean velocity (km/sec)
0.0–0.2	57
.2– .4	57
.4– .5	156
.5– .6	200
.6– .7	286

Now the fact that in our own neighborhood there must be members of the disk that are cluster-type variables was first shown in a very neat way by Struve. He discussed Joy's radial velocities for 133 cluster-type variables, essentially brighter than $12^m.5$, and found that the velocity relative to the sun has a remarkable dependence on the mean period (Table 24). The mean value of z for these 133 cluster-type variables is about 1.1 kpc, so they are pretty close to the plane of the Galaxy.

We know that a solution for the solar motion of all the cluster-type variables together always gives a very high value, usually of the order of 156 km/sec. But we see that the variables with the shortest periods give quite small velocities compared to the other groups. That must mean that the groups differ kinematically, although none of them is simon-pure. Now if you determine the velocity of the sun for disk stars at large, you get something of the order of 42 km/sec. And if you take the first two groups in the table you see that they give values fairly close to this. Since the rotational velocity of the Galaxy in the neighborhood of the sun is of the order of 216 km/sec, the value for a spherical distribution should be near to 200 km/sec. Our result clearly means that among the cluster-type variables there is a group that comes very near to the disk stars in velocity. As I said, this group may not be simon-pure yet; some of the stars should probably be rejected as outside the disk.

But the data show that among the cluster-type variables there is a group of true disk members, if we could only sort them out, and in doing so one should clearly give preference to the stars with periods shorter than $0^d.4$. I think no real attempt has yet been made to separate them, but I believe that Oosterhoff is right in suggesting that we should choose the cluster-type variables with periods shorter than $0^d.4$ that have asymmetric light curves and nearly normal amplitudes. The reason is that the cluster-type variables in the region of the galactic center, which have mostly periods shorter than $0^d.4$, have asymmetric light curves of large amplitude, not sine curves like the stars of the halo. This idea should be followed up by examining the light curves of Joy's stars: are they asymmetric? If some of them are, we have a chance right away of picking out cluster-type variables that are members of the disk. And the number in our neighborhood need not be especially high.

A third argument points in the same direction. When the distribution of cluster-type variables perpendicular to the galactic plane has been investigated, the data have usually been represented by a single straight line. But close examination of the data shows that we certainly have several curves with different gradients. For small values of z there is one group which gives rather a steep gradient, a second group which gives a smaller gradient, and a third group which is incomplete because of deficiencies in the data. When we come to within distances of 0.5 kpc of the plane, there is a clear-cut deficiency of observed stars, a result of absorption.

So there is not the slightest doubt that the disk of our Galaxy contains stars as old as the halo stars, and of the same composition as the halo stars. But in addition we have normal stars there, and we encounter the problem of the cause of the enrichment of old stars that set in very fast after star formation.

Let us turn now to the neighborhood of our sun, where we have excellent data so long as we restrict ourselves to the volume in which we have trigonometric parallaxes. We have to consider that our sun is located in a spiral arm — probably at the edge of it — in a region where star formation is going on. We know that our sun itself is an old star, 4.5×10^9 years old. A color-magnitude diagram for a given volume of space should give us information about the stars that are floating around in our neighborhood. In recent years we have obtained very well-determined photometric data on all the stars within 20 pc; we know that it provides a main sequence that includes such stars as Sirius and α Aquilae, and that the number of stars increases very rapidly as we go down. Actually a vast number of stars in such a volume are K dwarfs, with M dwarfs in even larger numbers.

For most of these stars we know the radial velocity, and the transverse velocities are quite well determined. The stars of the upper main sequence, say down to type G, have small space velocities; they are relatively younger stars, as was to be expected. But as soon as you come to classes later than G, the velocities alone show that you have a mixture of stars — very high velocities, low velocities, everything is represented. But the interesting thing is that in our neighborhood we get the first indication of stars that have turned off the main sequence at around $+3^M.5$ in appreciable numbers. Although the exact position of the turning-off branch is still poorly defined, simply because the individual parallaxes are sometimes not of high weight, there is not the slightest doubt that here we have old stars turning off the main sequence in the color-magnitude diagram. And the position of the branch leaves no doubt that we are dealing with old stars of the M 67 type. There may be a few stars of globular cluster type, halo type (ζ Herculis may be one), but their number must be exceedingly small. But there is no doubt about the other component; it contains old stars of metal-enriched type, and they have very moderate velocities, so they have really been in our neighborhood all the time and will stay there. A few stars lie outside even these limits, but I do not consider the data sufficient for us to jump to the conclusion that these are still older stars; we are getting only a first glimpse.

You see that our data about the disk of the Galaxy are exceedingly meager at present. Probably the next step will be the study of individual stars, for studies in the last few years have shown that interesting groups of stars can be separated out. Miss Roman, for instance, has found that she can divide the F stars in our neighborhood into strong-line and weak-line stars on the basis of their spectra. The next step showed that there is a very close correlation between these properties and the ultraviolet excesses of the stars: the extreme weak-line stars have large ultraviolet excesses, and Miss Roman found a continuous progression from these extreme stars to those that have essentially normal intensities.

Other methods have been developed in recent years for measuring parameters by which the position of a star can be checked, such as the method of measuring the intensity of the hydrogen lines and of the Balmer jump that is used by Barbier, Chalonge, and their group. Strömgren's photoelectric application of interference methods seems even more promising, and will lead very rapidly to an increase in our knowledge, especially of the masses and ages of the stars. A third type of investigation that has become possible recently is the use of the photoelectric scanner, first started by Whitford and Code. And all the methods can be used with instruments of moderate size.

I am very certain that work in this field will proceed rapidly. We shall no longer be restricted to star clusters, which after all provide a rather limited selection. Star clusters finally disintegrate, or become so small that there is nothing left, but the individual stars are still there.

This is all that I can say about the disk of the Galaxy; today we really know nothing. We only know that stars are present from the oldest to the youngest, and apparently of the most diversified chemical composition. But how they are arranged inside — that has still to be found out.

CHAPTER 23

KINEMATICS AND EVOLUTION

OF THE GALAXY

IT IS well known that the stars that make up the disk of our Galaxy show a variety of concentrations toward the galactic plane. The B and O stars, the Cepheids, and the gas are highly concentrated, whereas other disk groups, such as the novae and planetary nebulae, show considerably smaller concentration, not only toward the plane, but also toward the galactic nucleus.

For simplicity we shall consider just the concentration toward the galactic plane, which is the more important one in our neighborhood. It is clear that the concentrations of different groups of stars must be reflected in the motions of these stars, especially in the velocity dispersions $\sigma\hat{u}$, $\sigma\hat{v}$, $\sigma\hat{w}$, where u and v are the velocity components in the plane of the Galaxy, and w is that perpendicular to the plane. We know that our sun is in the galactic plane, and that all stars for which we can determine space motions (from radial velocities, proper motions, and distances) are very close to the sun, so that they also lie in or near the galactic plane. If a star reaches a large distance from the galactic plane, it has obviously to pass through the plane with rather high velocity. So stars with small concentration toward the plane will show a much larger value of σ_w than stars with a high concentration.

These relations between the distribution and motions of the stars were investigated in the 1920's by Stromberg, Lindblad, Oort, and others. Lindblad concluded that the Galaxy contains a number of interpenetrating subsystems. The idea of subsystems has been further developed by Kukarkin and Parenago. As a first step, Kukarkin selected well-defined groups of stars, such as certain groups of physical variables: cluster-type stars, long-period Cepheids, long-period variables, and novae. He investigated their spatial distribution and derived their galactic concentration in our neighborhood. He was able to represent the distributions by the exponential formula

$$\log N = a - b\,|z|,$$

where N is the number in a given volume of space and z is given in parsecs. The values of the coefficient b, which determines the concentration, are given in Table 25. He made no distinction between Cepheids of Type I and

Table 25. Galactic concentration of variable stars.

Type	b
Cepheids	0.00991
Novae	.0040:
Long-period, $P > 250^d$	.00084
Long-period, $P < 250^d$	.00028
Cluster-type stars	.00022

Type II; if he had done so, the value of b would have been even larger, because the concentration is very much larger for Cepheids of Type I. The value for novae is rather uncertain.

Although Kukarkin used the best available data, his coefficients are still very rough, first because the data are not as extensive as one would wish, and second because some of the groups are not homogeneous. I have mentioned the Cepheids; and he was obliged to divide the long-period variables into two groups. As I pointed out in the last chapter, the cluster-type variables also consist of two groups, one of which is concentrated to the disk. If he had included the globular clusters, he would have got a still smaller value of b for them, but they also are a mixed group including a small number of disk clusters. In general, Kukarkin was very careful, and he provided data about the galactic concentrations that should not be seriously off.

The next step, taken by Parenago, was to investigate the same groups of stars kinematically, and to derive the velocity dispersions σ_u, σ_v, σ_w, for them. He showed that there was a close correlation between distribution and kinematic properties. Accordingly Kukarkin and Parenago were able to divide their material roughly into four groups (Table 26); there are continuous transitions between groups.

The first group contains rather well-defined objects: gas, which is very strongly concentrated; B stars; Cepheids (young stars, which should be concentrated with the gas); supergiants (also all young stars); and open clusters (at least the young ones). We know today that there are some old objects among the open clusters, although young ones predominate. An old cluster like M 67 is 400 pc above the plane, so one has to be careful. One would like the

Table 26.   Subsystems of the galactic disk.

Subsystem	Components
1. Very flat	Interstellar gas; B stars; Cepheids; supergiants; open clusters
2. Flat	A stars; giant stars from G to M; R and N stars
3. Intermediate	Long-period variables $(P > 250^{\mathrm{d}})$; subgiants; white dwarfs; planetary nebulae; dwarf stars, G to M
4. Spherical	Long-period variables $(P < 250^{\mathrm{d}})$; subdwarfs; high-velocity giants; cluster-type variables; globular clusters

open clusters to be investigated according to their ages, but the table gives a first approximation, which is reasonable because most open clusters are young.

The novae are not included in the table, because, although we can determine their concentration to the plane, their velocities are very difficult to determine. Their concentration would put them with the A stars, but they are certainly very much more widely scattered, and should probably go in the third group. The A stars certainly belong in the Flat group.

It is clear that Kukarkin and Parenago followed out the general idea of Lindblad's subsystems, and arranged the stars in a sequence of well-defined groups, with the expectation that the result would in some way reflect the history of our Galaxy. But as it stands their arrangement does not involve any dating (unless additional assumptions are made), and does not even indicate the direction in which development has run. Both dating and direction must be approached in another way.

I do not think that this is an overstatement, because,

as I see it, the most highly concentrated systems are the youngest. Kukarkin, on the other hand, would like to have it run the other way, because he believed recently that star formation is going on in globular clusters. The arrangement of Table 26 is a formal arrangement, and gives no information about ages or direction of development; these must be provided by a study of physical properties.

I think that we can approach the problem by selecting stars that are of the same age, or groups of stars with a very limited range of age. Today we certainly understand one feature of the picture, the reason why the youngest stars that we know — the O and B stars and the Cepheids, are so strongly concentrated to the plane of the Galaxy. It is because the gas is concentrated to such a remarkable degree, and the young stars still reflect the kinematical properties of the gas in which they were formed.

Now it is intriguing to see whether we can follow this idea up by working with different age groups, and trace the history of a subsystem, which we can call gas that may originally have extended throughout the whole disk, and grew more concentrated by and by. This, of course, is a hypothesis that must be followed up, and, since the oldest stars will be distributed throughout the whole Galaxy, it will be quite a job. The most direct approach will again be through the spatial distribution. We should select the old stars, which can easily be recognized nowadays; the multicolor photometry of Strömgren will certainly open up the way to do it. The whole analysis should now be repeated with well-defined age groups, to see what comes out. Well, that is the music of the future.

Although kinematics alone cannot furnish the time data, in special cases it can provide quite interesting information. Let me take one example. In the disk in the neighborhood of our sun we have a mixture of young and old stars. If we consider only the main sequence, the youngest stars would run all the way down from the highest luminosities; they would show a very strong concentration to the plane of the Galaxy, and their velocity dispersion in the perpendicular direction would be very small. On the other hand, in the extreme group of the oldest stars, the main sequence begins only at $+3^M.5$; they will have low concentration to the galactic plane and larger velocity dispersion. These are the two extremes, and between them will lie all the intermediate cases, from young to old.

It is clear that, if we study a group of stars of all ages in the neighborhood of the sun, the kinematic properties of the stars in the upper main sequence will approach those of young stars, whereas those of the lower main sequence will come from a mixture of stars of all ages. The situation is rather complicated, because our sun is located at the edge of a spiral arm. What is the ratio of old stars to young stars in our neighborhod? The facts indicate quite clearly that, even though we are in a spiral arm, star formation is still going on in our neighborhood. When we go to the dwarf stars we find that the old stars absolutely dominate.

The first investigation of the kinematics of main-sequence stars alone was made a few years ago by Parenago. It is rather simple to sort out all the stars that are not on the main sequence; occasionally a subgiant or even a subdwarf might slip in, but Parenago made a special search for subgiants. He used quite good data, and his list should be simon-pure. He plotted the solar velocity referred to each spectral group from B to M. All the youngest stars, up to F, give more or less the same solar velocity. But between F and G, in the region of $+3^M.5$, there is a sudden jump in solar velocity.

Now the region in which the oldest stars turn off the main sequence is between F and G at about $+3^M.5$. In principle, the G and K stars could be a straight mixture of old and young stars, and the solar velocity in this part of the diagram could be a weighted mean for old and young stars. But the subgiants are only old stars, and the fact that they lie near the line for the G, K, M stars shows that this line gives essentially the velocity of old stars.

Similarly, the velocity dispersion for these groups of stars shows a sudden break between F and G at $+3^M.5$, and again the subgiants fit into the picture. So we conclude that although in our neighborhood we have a mixture of old and young stars, the overwhelming number of the fainter stars are old; they are actually among the oldest stars, because the subgiants are among the oldest stars. The further conclusion is that most of the stars in our Galaxy must have been formed right at the beginning, say in the first few billion years. There was a big burst of star formation, and afterward the rate of formation must have dropped off very fast, according to these data.

In order to get even more precise data, Parenago went beyond the trigonometric parallaxes, and used the spectroscopic parallaxes for stars nearer than 20 pc, taken from the very careful list recently prepared by Gliese. He now studied the space motions, which should reflect the same effect. There are no B stars within 20 pc, and we start with A5 stars, for which the mean space velocity is about 19 km/sec. The relation between spectral type and space velocity gives us the same story again.

These data are extremely interesting, because they show that as we go to smaller masses — say stars fainter than $+4^M$ — the majority of the stars in our neighborhood are old stars, although we are in a spiral arm.

Our sun is 4.5×10^9 years old, and its velocity relative to the A stars or the B stars is 20 km/sec. Its velocity vector points inward compared to the circular vector, which places it definitely with the older stars of higher velocity dispersion. Vyssotsky has shown that there is a good indication that our sun should be counted among the weak-line stars, which are not quite up to the normal metal content as shown by the interstellar gas or the present O and B stars. If we want to assign the sun to one of the two populations, it depends on what we prefer as a classification argument — age or chemical composition or a combination of the two. My personal feeling is that the age is the most interesting thing for over-all purposes in our Galaxy. We could consider enrichment as a concomitant thing. As to age, we have a continuous grouping: star formation is going on all the time. But the fact that apparently most of the stars in our neighborhood must have been formed at a very early time makes it possible, I think, to speak in a rough sense of two populations, one corresponding to this big burst at the beginning, the other to what you might call the tail end. Probably, since the sun is 4.5 billion years old, you could assign it to either population, perhaps late Population II or early Population I.

The discontinuity about which I have spoken is indicated by a number of other things; the old group of stars preponderates. For instance, the transverse motions of main-sequence stars show a velocity dispersion that cannot be represented by a single function, but by two with different dispersions; the final results are similar to Parenago's.

Recently Vyssotsky has derived the solar velocities for Miss Roman's two groups of strong-line and weak-line stars, both for his own data and for hers. The strong-line group contains the younger stars. You see that the result is always the same (Table 27).

Table 27. Solar velocity for strong-line and weak-line stars.

Group	Miss Roman's data (km/sec)	Vyssotsky's data (km/sec)
Strong-line	13 ± 3	15
Weak-line	21 ± 4	21

This result brings up another question, to which we do not know the answer: has the mass spectrum of the stars that have been formed always been the same? Were masses formed with the same frequency in early times as they are today? Doubts about the constancy of the mass function arose rather early, for instance when Hertzsprung showed many years ago that the luminosity function of the Pleiades is deficient in small masses. Conversely, Oort showed in his investigation of the high-velocity stars, which represent the population at the center of the Galaxy, that, compared to the number of F stars on the main sequence, the percentage of G, K, and M dwarfs in the group is very much higher than the general luminosity function would indicate.

The luminosity function in our Galaxy is still quite uncertain, and it is not a unique thing, but is made up from all kinds of stars, formed at all epochs. But the problem becomes far simpler now that we know that there was probably a big burst of star formation everywhere right at the beginning. Maybe in the first approximation we can just consider two phases of star formation. The question then is whether the mass spectrum was the same for both. The early work, and especially Oort's investigation of the high-velocity stars, made this seem very doubtful.

The presently accepted luminosity function is the van Rhijn function supplemented by the data of Luyten and others; we know that it must extend to $+20^M$. We are now interested in the faint stars. The old Kapteyn function of 1905 had a maximum at about $+7$, but the new data of the 1920's on proper-motion stars suddenly extended the curve and pushed the maximum to a much fainter luminosity. In a few years, when the plates of the Palomar Sky Survey have been repeated, we shall have a much better luminosity function for stars fainter than $+12^M$; the Schmidt is an ideal instrument for the purpose, and we shall have proper-motion data down to the twentieth magnitude.

Seares made a very interesting discussion of the part of the luminosity-function that had been contributed by the proper motions discovered by Max Wolf, Ross, and others. He asked and answered the very interesting question: how do the kinematic properties of the $+15^M$ stars compare with those of the $+5^M$ stars? As he had only proper motions available, he used the invariant $\overline{h} = m + 5 \log \mu$, which is independent of the distance of the star. Seares showed conclusively that the mean transverse motion for the fainter stars was 1.7 times that for the brighter stars. If this is so, it means that the fainter stars have a higher velocity dispersion.

We should not forget that the data for the brighter stars refer to a rather small region about the sun, and that this region is more restricted the fainter we go. So the difference in velocity dispersion is really significant. From the looks of it, I should not be surprised if it turned out that the faint part of the luminosity function is essentially contributed by other older stars, the other part by younger ones. This would be in agreement with Oort's result, that the numbers of main-sequence stars of a given spectral type increase very much faster at faint absolute magnitudes for the high-velocity stars than for the normal stars.

It will be interesting to find out whether we are really dealing with the superposition of two luminosity functions, which would mean two mass spectra. At present the data are suggestive, but not conclusive. I expect we shall have the answer within the next 12 or 15 years, and I think it will probably be a very important one.

One thing only we can say about the mass spectrum. Elliptical galaxies belong to the group where one would expect the higher frequency of small masses. And we should expect a low frequency of small masses in condensed strips, such as the thin central layer of our Galaxy. We can hardly guess what causes the difference; it probably lies in the mechanism and process of star formation.

This brings me to the end of my lectures. I hope I have brought home sufficiently how little we know, and how large the gaps are. In many ways I envy those who will have to fill the gaps. I only wish that I were young, and could start all over again.

INDEX